Real-World Data Mining

Books in the FT Press Analytics Series

Enterprise Analytics by Thomas Davenport and the International Institute for Analytics (ISBN: 0133039439)

People Analytics by Ben Waber (ISBN: 0133158314)

A Framework for Applying Analytics in Healthcare by Dwight McNeill (ISBN: 0133353745)

Modeling Techniques in Predictive Analytics by Thomas W. Miller (ISBN: 0133412938)

Applying Advanced Analytics to HR Management Decisions by James Sesil (ISBN: 0133064603)

The Applied Business Analytics Casebook by Matthew Drake (ISBN: 0133407365)

Analytics in Healthcare and the Life Sciences by Thomas Davenport, Dwight McNeill, and the International Institute for Analytics (ISBN: 0133407330)

Managerial Analytics by Michael Watson and Derek Nelson (ISBN: 013340742X)

Data Analytics for Corporate Debt Markets by Robert S. Kricheff (ISBN: 0133553655)

Business Analytics Principles, Concepts, and Applications by Marc J. Schniederjans, Dara G. Schniederjans, and Christopher M. Starkey (ISBN: 0133552187)

Big Data Analytics Beyond Hadoop by Vijay Agneeswaran (ISBN: 0133837947)

Computational Intelligence in Business Analytics by Les Sztandera (ISBN: 013355208X)

Big Data Driven Supply Chain Management by Nada R. Sanders (ISBN: 0133801284)

Marketing and Sales Analytics by Cesar Brea (ISBN: 0133592928)

Cutting-Edge Marketing Analytics by Rajkumar Venkatesan, Paul Farris, and Ronald T. Wilcox (ISBN: 0133552527)

Applied Insurance Analytics by Patricia L. Saporito (ISBN: 0133760367)

Modern Analytics Methodologies by Michele Chambers and Thomas W. Dinsmore (ISBN: 0133498581)

Advanced Analytics Methodologies by Michele Chambers and Thomas W. Dinsmore (ISBN: 0133498603)

Modeling Techniques in Predictive Analytics, Revised and Expanded Edition, by Thomas W. Miller (ISBN: 0133886018)

Modeling Techniques in Predictive Analytics with Python and R by Thomas W. Miller (ISBN: 0133892069)

Business Analytics Principles, Concepts, and Applications with SAS by Marc J. Schniederjans, Dara G. Schniederjans, and Christopher M. Starkey (ISBN: 0133989402)

Profiting from the Data Economy by David A. Schweidel (ISBN: 0133819779)

Business Analytics with Management Science Models and Methods by Arben Asllani (ISBN: 0133760359)

Digital Exhaust by Dale Neef (ISBN: 0133837963)

Web and Network Data Science by Thomas W. Miller (ISBN: 0133836441)

Applied Business Analytics by Nathaniel Lin (ISBN: 0133481506)

Trends and Research in the Decision Sciences by Decision Sciences Institute and Merrill Warkentin (ISBN: 0133925374)

Real-World Data Mining

Applied Business Analytics and Decision Making

Dursun Delen, Ph.D.
Professor of Management Science and Information Systems
Spears and Patterson Endowed Chairs in Business Analytics
Spears School of Business, Oklahoma State University

Editor-in-Chief: Amy Neidlinger
Executive Editor: Jeanne Glasser Levine
Development Editor: Natasha Wolmers
Operations Specialist: Jodi Kemper
Cover Designer: Alan Clements
Managing Editor: Kristy Hart
Senior Project Editor: Betsy Gratner
Copy Editor: Kitty Wilson
Proofreader: Kathy Ruiz
Indexer: Tim Wright
Compositor: Nonie Ratcliff
Manufacturing Buyer: Dan Uhrig

Published by Pearson Education, Inc.
Upper Saddle River, New Jersey 07458

For information about buying this title in bulk quantities, or for special sales opportunities (which may include electronic versions; custom cover designs; and content particular to your business, training goals, marketing focus, or branding interests), please contact our corporate sales department at corpsales@pearsoned.com or (800) 382-3419.

For government sales inquiries, please contact governmentsales@pearsoned.com.

For questions about sales outside the U.S., please contact international@pearsoned.com.

Printed in the United States of America

First Printing December 2014

ISBN-10: 0-13-355107-5
ISBN-13: 978-0-13-355107-5

Pearson Education LTD.
Pearson Education Australia PTY, Limited.
Pearson Education Singapore, Pte. Ltd.
Pearson Education Asia, Ltd.
Pearson Education Canada, Ltd.
Pearson Educación de Mexico, S.A. de C.V.
Pearson Education—Japan
Pearson Education Malaysia, Pte. Ltd.

Library of Congress Control Number: 2014953043

To my mother;
without her love and compassion,
I wouldn't be here.

Contents

Foreword

Dr. Dursun Delen has written a concise, information-rich book that effectively provides an excellent learning tool for someone who wants to understand what analytics, data mining, and "Big Data" are all about. As business becomes increasingly complex and global, decision-makers must act more rapidly and accurately, based on the best available evidence. Modern data mining and analytics are indispensable for doing this. This reference makes clear the current best practices, demonstrates to readers—students and practitioners—how to use data mining and analytics to uncover hidden patterns and correlations, and explains how to leverage these to improve decision-making.

The author delivers the right amount of concept, technique, and cases to help readers truly understand how data mining technologies work. Coverage includes data mining processes, methods, and techniques; the role and management of data; tools and metrics; text and web mining; sentiment analysis; and integration with cutting-edge Big Data approaches, as presented as follows.

In Chapter 1, he commendably traces the roots of analytics from World War II times to the present, illustrated by Figure 1.2, where he takes the reader from *Decision Support Systems* in the 1970s, to the *Enterprise/Executive IS Systems* in the 1980s, to the *Business Intelligence (BI)* that we all heard about in the 1990s and early 2000s, and finally to our modern day uses of *analytics* (2000s) and *Big Data* (2010s). That was all in Chapter 1, creating a preamble for what is to come in the rest of the book: data mining.

Chapter 2 provides a very easy-to-understand description and an excellent taxonomy for data mining. In this chapter, the author differentiates data mining from several other related terminologies, making a strong case for what it really stands for: discovery of knowledge. Identifying data mining as a problem-solving and decision-making philosophy that sits in the midst of many disciplines is quite refreshing; many people think of data mining as a new discipline of its own. With a number of real-world examples, intuitive graphics, and down-to-earth discussion, this chapter demystifies data mining for the masses. In my opinion, this is an excellent way to portray seemingly complex and highly technical topics like data mining to a wider audience.

In Chapter 3, Dr. Delen provides a rich collection of different approaches to standardized data mining processes in a manner that any reader can understand. KDD (knowledge discovery in databases) is the first standardized process that the chapter talks about, which was developed by Usama Fayyad, an early pioneer in the field. Dr. Delen presents KDD in an engaging discussion enhanced with a diagram (Figure 3.1), which illustrates the flow of the KKD data mining process. Additionally, other data mining "schemas" proposed by various groups and individuals are examined to show the development of the fundamental thinking in this field. To illustrate the usefulness of these schemas, Dr. Delen presents a data mining case study at the end of this chapter: "Mining Cancer Data for New Knowledge."

Chapter 4 considers the types of data used in data mining, including the ever-increasing use of text data (that is, unstructured non-numerical data, which is probably 90% of the data available to the world today). Data preparation is the most important part of data mining: Data must be clean and good in order to develop useful models ("garbage in, garbage out"); thus, up to 90% of the time involved in data mining can be taken up by the data preparation stage. Dr. Delen goes into all the ways of looking at data to get it clean and ready for data analytics, including developing the train and test data sets, giving one of the most learner-friendly visuals of *k*-fold cross-validation in Figure 4.6.

In Chapter 5, Dr. Delen describes the most common data mining algorithms in a way that the layperson can understand. Among others, neural networks and SVM (support vector machines) are described thoroughly, with illustrations that help the reader really understand these complicated mathematical processes. Dr. Delen makes his own original illustrations, and they are well worth the price of the book!

Text mining (text analytics) is described thoroughly in Chapter 6, with Dr. Delen starting out with a diagram he originally made for our 2012 book, *Practical Data Mining* (on which I am the lead author: Miner, G.D.; Delen, D.; Elder, J.; Fast, A.; Hill, T.; and Nisbet, B. Elsevier/Academic Press: 2012). Dr. Delen effectively distills our large 1,100 page book into one chapter that tells it all—in other words, is very useful for the new learner. Well done!

In the last chapter, Chapter 7, Dr. Delen goes into the new buzz word in the analytics field: Big Data analytics. Big Data is heard in the news almost daily. What does it mean? It means different things to different people. But I can tell you that, working in the data mining field for more than 15 years now, I have been dealing with Big Data all that time. But the ever-decreasing cost of storage space for data and the availability of cloud storage, plus the availability of faster and faster computers, mean that even a small laptop can do both distributed processing and multi-threading in data analysis. This has made even the small tablet more powerful than the warehouse of air-conditioned mainframe servers of decades ago. One can even run a bank of servers and cloud storage from one's smartphone these days. So as the data becomes "bigger," the physical needs to process it become "smaller."

But Big Data is misunderstood by most, at least it seems that way to me. Many think that data mining requires Big Data. But I have worked with medical residents for 10 years who want to look at lots of variables in their one-year research project but usually can only get a fraction of the cases they need for that many variables in their limited time; thus, traditional statistics are of almost no use to these paltry data sets by traditional statistics standards; yet by using machine-learning, modern data mining methods, I have found that one can usually generate useful hypotheses from these small data sets and find knowledge that was previously impossible to obtain with only traditional *p*-value Fischerian statistics. Traditional statistics were an anomaly of the twentieth century; prior to 1900, Bayesian statistics had predominated in data analysis for centuries; with the advent of the year 2000, the new modern versions of Bayesian statistics—including the SVM, NN, and other machine-learning modalities—had come of age, and we are now back into the Bayesian age in this twenty-first century. Unfortunately, it is taking a while for the "traditionally statistical training" cadre to understand and catch up…but the cutting edge is with Bayesian, data mining, and Big Data.

Anyone wanting to learn about data mining and have a technical understanding of the topic should get this book. By the end of the read, you will understand the field!

—Gary D. Miner, Ph.D.
Author of two PROSE Award–winning analytics books
Senior Analyst, Healthcare Applications Specialist
Dell, Information Management Group, Dell Software

Acknowledgments

Without a doubt, we are living in the age of data mining and Big Data analytics. Because of their popularity (and perhaps a little bit of hype), everybody is talking about data mining and Big Data analytics, often in different scope and contexts, using diverse terminology. The main goal of this book is to explain the language of analytics and data mining in a comprehensive yet not-too-technical way. If I have, at least partially, succeeded in achieving this goal, it is because of the direct and indirect contributions of a number of people.

I want to thank my colleagues and my students for providing me with the broader perspective toward analytics that I needed to write this book in a holistic manner. As is the case for most academics, I also have my own opinions and biases toward "what is what" in data mining and in analytics, and thanks to my academic friends, I think I managed to rise above them to make this book inclusive and comprehensive.

We academics tend to be focused on rigor and theory, sometimes in the process moving away from relevance and reality. Thanks to my clients and corporate partners who continuously provide me with the realities of the real world that I need to stay balanced between rigor and relevance. Writing a book titled *Real-World Data Mining* requires such connection to reality, without compromising technical accuracy, and I have to thank to my corporate friends for helping me achieve that in this book.

I want to thank to my publisher, Ms. Jeanne Levine, and Pearson for presenting me with the opportunity to write this book and for being patient and resourceful for me throughout the journey of actually writing it.

About the Author

Dr. Dursun Delen is an internationally known expert in business analytics and data mining. He is often invited to national and international conferences to deliver keynote presentations on topics related to data/text mining, business intelligence, decision support systems, business analytics, and knowledge management. Prior to his appointment as professor at Oklahoma State University in 2001, Dr. Delen worked for industry for more than 10 years, developing and delivering business analytics solutions to companies. Most recently he worked for a privately owned research and consulting company, Knowledge Based Systems, Inc., in College Station, Texas, as a research scientist. During his five years there, he led a number of projects related to decision support, information systems, and advanced analytics that were funded by federal agencies, including the DoD, NASA, NIST, and the DOE. Today, in addition to his academic endeavors, Dr. Delen provides consulting services to businesses in assessing their information systems needs and developing state-of-the-art business analytics capabilities.

Dr. Delen holds William S. Spears and Neal Patterson Endowed Chairs in Business Analytics, and he is Director of Research for the Center for Health Systems Innovation and Professor of Management Science and Information Systems in the Spears School of Business at Oklahoma State University. His research has appeared in major journals, including *Decision Sciences*, *Decision Support Systems*, *Communications of the ACM*, *Computers and Operations Research*, *Computers in Industry*, *Journal of Production Operations Management*, *Artificial Intelligence in Medicine*, and *Expert Systems with Applications*, among others. He has recently published six books: *Advanced Data Mining Techniques* (Springer, 2008), *Decision Support and Business Intelligence Systems* (Prentice Hall, 2010), *Business*

Intelligence: A Managerial Approach (Prentice Hall, 2010), *Practical Text Mining and Statistical Analysis for Non-structured Text Data Applications* (Elsevier, 2012), *Business Intelligence: A Managerial Perspective on Analytics*, 3rd edition (Prentice Hall, 2013), and *Business Intelligence and Analytics: Systems for Decision Support*, 10th edition (Prentice Hall, 2014). He served as the general co-chair for the fourth International Conference on Network Computing and Advanced Information Management, and he regularly chairs tracks and mini-tracks at various information systems conferences. He also serves as associate editor-in-chief, senior editor, associate editor, and editorial board member on a dozen academic and technical journals.

1

Introduction to Analytics

Business analytics is a relatively new term that is gaining popularity in the business world like nothing else in recent history. In general terms, *analytics* is the art and science of discovering insight—by using sophisticated mathematical models along with a variety of data and expert knowledge—to support solid, timely decision making. In a sense, analytics is all about decision making and problem solving. These days, *analytics* can be defined as simply as "the discovery of meaningful patterns in data." In this era of abundant data, analytics tends to be used on large quantities and varieties of data. Although analytics tends to be data focused, many applications of analytics involve very little or no data; instead, those analytics projects use mathematical models that rely on process description and expert knowledge (e.g., optimization and simulation models).

Business analytics is the application of the tools, techniques, and principles of analytics to complex business problems. Firms commonly apply analytics to business data to describe, predict, and improve business performance. Firms have used analytics in many ways, including the following:

- To improve their relationships with their customers (encompassing all phases of customer relationship management—acquisition, retention, and enrichment), employees, and other stakeholders

- To identify fraudulent transactions and odd behaviors—and, in doing so, saving money
- To enhance product and service features and their pricing, which would lead to better customer satisfaction and profitability
- To optimize marketing and advertising campaigns so they can reach more customers with the right kind of message and promotions with the least amount of expense
- To minimize operational costs by optimally managing inventories and allocating resources wherever and whenever they are needed by using optimization and simulation modeling
- To empower employees with the information and insight they need to make faster and better decisions while they are working with customers or customer-related issues

The term *analytics*, perhaps because of its rapidly increasing popularity as a buzzword, is being used to replace several previously popular terms, such as *intelligence*, *mining*, and *discovery*. For example, the term *business intelligence* has now become *business analytics*; *customer intelligence* has become *customer analytics*, *Web mining* has become *Web analytics*, *knowledge discovery* has become *data analytics*, etc. Modern-day analytics can require extensive computation because of the volume, variety, and velocity of data (which we call *Big Data*). Therefore, the tools, techniques, and algorithms used for analytics projects leverage the most current, state-of-the-art methods developed in a variety of fields, including management science, computer science, statistics, data science, and mathematics. Figure 1.1 shows a word cloud that includes concepts related to analytics and Big Data.

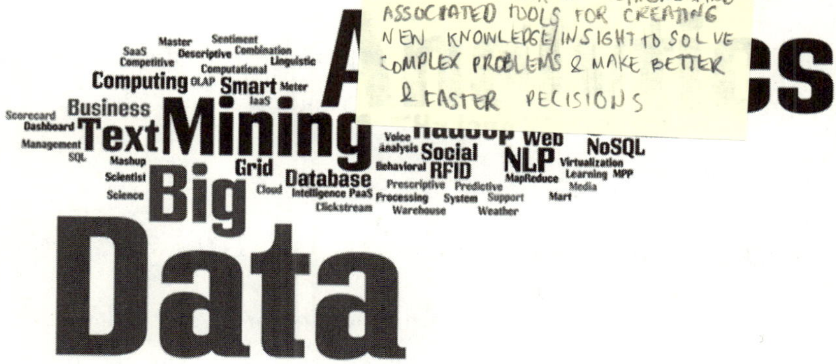

Figure 1.1 Analytics and Big Data Word Cloud

Is There a Difference Between Analytics and Analysis?

Even though the two terms *analytics* and *analysis* are often used interchangeably, they are not the same.

Basically, *analysis* refers to the process of separating a whole problem into its parts so that the parts can be critically examined at the granular level. It is often used when the investigation of a complete system is not feasible or practical, and the system needs to be simplified by being decomposed into more basic components. Once the improvements at the granular level are realized and the examination of the parts is complete, the whole system (either a conceptual or physical system) can then be put together using a process called *synthesis*.

Analytics, on the other hand, is a variety of methods, technologies, and associated tools for creating new knowledge/insight to solve complex problems and make better and faster decisions. In essence, analytics is a multifaceted and multidisciplinary approach to addressing complex situations. Analytics take advantage of data and mathematical models to make sense of the complicated world we are living in. Even though analytics includes the act of analysis at different stages of the discovery process, it is not just analysis but also includes synthesis and other complementing tasks and processes.

Where Does Data Mining Fit In?

Data mining is the process of discovering new knowledge in the forms of patterns and relationships in large data sets. The goal of analytics is to convert data/facts into actionable insight, and data mining is the key enabler of that goal. Data mining has been around much longer than analytics, at least in the context of analytics today. As *analytics* became an overarching term for all decision support and problem-solving techniques and technologies, data mining found itself a rather large space within that arc, ranging from descriptive exploration of identifying relationships and affinities among variables (e.g., market-basket analysis) to developing models to estimate future values of interesting variables. As we will see later in this chapter, within the taxonomy of analytics, data mining plays a key role at every level, from the most simple to the most sophisticated.

Why the Sudden Popularity of Analytics?

Analytics is a buzzword of business circles today. No matter what business journal or magazine you look at, it is very likely that you will see articles about analytics and how analytics is changing the way managerial decisions are being made. It has become a new label for evidence-based management (i.e., evidence/data-driven decision making). But why has analytics become so popular? And why now? The reasons (or forces) behind this popularity can be grouped into three categories: need, availability and affordability, and culture change.

Need

As we all know, business is anything but "as usual" today. Competition has been characterized progressively as local, then regional, then national, but it is now global. Large to medium to small, every business is under the pressure of global competition. The tariff and transportation cost barriers that sheltered companies in their respective geographic locations are no longer as protective as they once were. In addition to (and perhaps because of) the global competition, customers have become more demanding. They want the highest quality of products and/or services with the lowest prices in the shortest possible time. Success or mere survival depends on businesses being agile and their managers making the best possible decisions in a timely manner to respond to market-driven forces (i.e., rapidly identifying and addressing problems and taking advantage of the opportunities). Therefore, the need for fact-based, better, and faster decisions is more critical now than ever before. In the midst of these unforgiving market conditions, analytics is promising to provide managers the insights they need to make better and faster decisions, which help improve their competitive posture in the marketplace. Analytics today is widely perceived as saving business managers from the complexities of global business practices.

Availability and Affordability

Thanks to recent technological advances and the affordability of software and hardware, organizations are collecting tremendous amounts of data. Automated data collections systems—based on a variety of sensors and RFID—have significantly increased the quantity and quality of organizational data. Coupled with the content-rich data collected from Internet-based technologies such as social media, businesses now tend to have more data than they can handle. As the saying goes, "They are drowning in data but starving for knowledge."

Along with data collection technologies, data processing technologies have also improved significantly. Today's machines have numerous processors and very large memory capacities, so they are able to process very large and complex data in a reasonable time frame—often in real time. The advances in both hardware and software technology are also reflected in the pricing, continuously reducing the cost of ownership for such systems. In addition to the ownership model, along came the software- (or hardware-) as-a-service business model, which allows businesses (especially small to medium-size businesses with limited financial power) to rent analytics capabilities and pay only for what they use.

Culture Change

At the organizational level, there has been a shift from old-fashioned intuition-driven decision making to new-age fact-/evidence-based decision making. Most successful organizations have made a conscious effort to shift to data-/evidence-driven business practices. Because of the availability of data and supporting IT infrastructure, such a paradigm shift is taking place faster than many thought it would. As the new generation of quantitatively savvy managers replaces the baby boomers, this evidence-based managerial paradigm shift will intensify.

The Application Areas of Analytics

Even though the business analytics wave is somewhat new, there are numerous applications of analytics, covering almost every aspect of business practice. For instance, in customer relationship management, a wealth of success stories tell of sophisticated models developed to identify new customers, look for up-sell/cross-sell opportunities, and find customers' with a high propensity toward attrition. Using social media analytics and sentiment analysis, businesses are trying to stay on top of what people are saying about their products/services and brands. Fraud detection, risk mitigation, product pricing, marketing campaign optimization, financial planning, employee retention, talent recruiting, and actuarial estimation are among the many business applications of analytics. It would be very hard to find a business issue where a number of analytics application could not be found. From business reporting to data warehousing, from data mining to optimization analytics, techniques are used widely in almost every facet of business.

The Main Challenges of Analytics

Even though the advantages of and reasons for analytics are evident, many businesses are still hesitant to jump on the analytics bandwagon. These are the main roadblocks to adoption of analytics:

- **Analytics talent.** Data scientists, the quantitative geniuses who can convert data into actionable insight, are scarce in the market; the really good ones are very hard to find. Because analytics is relatively new, the talent for analytics is still being developed. Many colleges have started undergraduate and graduate programs to address the analytics talent gap. As the popularity of analytics increases, so will the need for people who have the knowledge and skills to convert Big Data into information and knowledge that managers and other decision makers need to tackle real-world complexities.
- **Culture.** As the saying goes, "Old habits die hard." Changing from a traditional management style (often characterized by intuition as the basis of making decision) to a contemporary management style (based on data and scientific models for managerial decisions and collective organizational knowledge) is not an easy process to undertake for any organization. People do not like to change. Change means losing what you have

learned or mastered in the past and now needing to learn how to do what you do all over again. It suggests that the knowledge (which is also characterized as power) you've accumulated over the years will disappear or be partially lost. The culture shift may be the most difficult part of adopting analytics as the new management paradigm.

- **Return on investment.** Another barrier to adoption of analytics is the difficulty in clearly justifying its return on investment (ROI). Analytics projects are complex and costly endeavors, and their return is not immediately clear, many executives are having a hard time investing in analytics, especially on large scales. Will the value gained from analytics outweigh the investment? If so, when? It is very hard to convert the value of analytics into justifiable numbers. Most of the value gained from analytics is somewhat intangible and holistic. If done properly, analytics could transform an organization, putting it on a new and improved level. A combination of tangible and intangible factors needs to be brought to bear to numerically rationalize investment and movement toward analytics and analytically savvy management practice.
- **Data.** The media is taking about Big Data in a very positive way, characterizing it as an invaluable asset for better business practices. This is mostly true, especially if the business understands and knows what to do with it. For those who have no clue, Big Data is a big challenge. Big Data is not just big; it is unstructured, and it is arriving at a speed that prohibits traditional collection and processing means. And it is usually messy and dirty. For an organization to succeed in analytics, it needs to have a well-thought-out strategy for handling Big Data so that it can be converted to actionable insight.

- **Technology.** Even though technology is capable, available, and, to some extent, affordable, technology adoption poses another challenge for traditionally less technical businesses. Although establishing an analytics infrastructure is affordable, it still costs a significant amount of money. Without financial means and/or a clear return on investment, management of some businesses may not be willing to invest in needed technology. For some businesses, an analytics-as-a-service model (which includes both software and the infrastructure/hardware needed to implement analytics) may be less costly and easier to implement.
- **Security and privacy**. One of the most common criticisms of data and analytics is the security. We often hear about data breaches of sensitive information, and indeed, the only completely secured data infrastructure is isolated and disconnected from all other networks (which goes against the very reason for having data and analytics). The importance of data security has made information assurance one of the most popular concentration areas in information systems departments around the world. At the same time that increasingly sophisticated techniques are being used to protect the information infrastructure, increasingly sophisticated attacks are becoming common. There are also concerns about personal privacy. Use of personal data about customers (existing or prospective), even if it is within legal boundaries, should be avoided or carefully scrutinized to protect an organization against bad publicity and public outcry.

Despite the hurdles in the way, analytics adoption is growing, and analytics is inevitable for today's enterprises, regardless of size or industry segment. As the complexity in conducting business increases, enterprises are trying to find order in the midst of the chaotic behaviors. The ones that succeed will be the ones fully leveraging the capabilities of analytics.

A Longitudinal View of Analytics

Although the buzz about it is relatively recent, analytics isn't new. It's possible to find references to corporate analytics as far back as the 1940s, during the World War II era, when more effective methods were needed to maximize output with limited resources. Many optimization and simulation techniques were developed then. Analytical techniques have been used in business for a very long time. One example is the time and motion studies initiated by Frederick Winslow Taylor in the late 19th century. Then Henry Ford measured pacing of assembly lines, which led to mass-production initiatives. Analytics began to command more attention in the late 1960s, when computers were used in decision support systems. Since then, analytics has evolved with the development of enterprise resource planning (ERP) systems, data warehouses, and a wide variety of other hardware and software tools and applications.

The timeline in Figure 1.2 shows the terminology used to describe analytics since the 1970s. During the early days of analytics, prior to the 1970s, data was often obtained from the domain experts using manual processes (i.e., interviews and surveys) to build mathematical or knowledge-based models to solve constraint optimization problems. The idea was to do the best with limited resources. Such decision support models were typically called *operations research* (*OR*). The problems that were too complex to solve optimally (using linear or non-linear mathematical programming techniques) were tackled using heuristic methods such as simulation models.

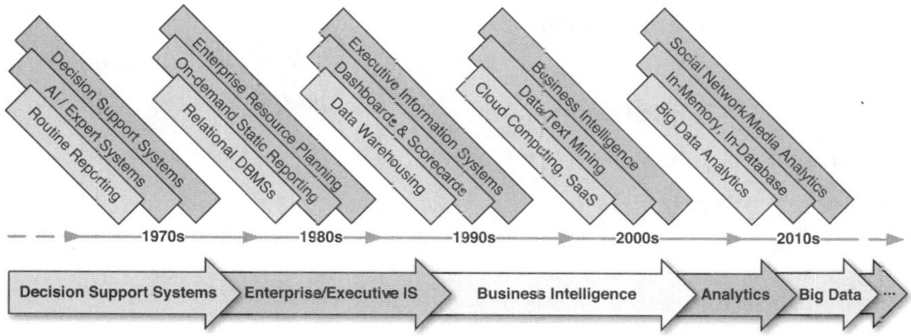

Figure 1.2 A Longitudinal View of the Evolution of Analytics

In the 1970s, in addition to the mature OR models that were being used in many industries and government systems, a new and exciting line of models had emerged: rule-based expert systems (ESs). These systems promised to capture experts' knowledge in a format that computers could process (via a collection of if–then rules) so that they could be used for consultation much the same way that one would use domain experts to identify a structured problem and to prescribe the most probable solution. ESs allowed scarce expertise to be made available where and when needed, using an "intelligent" decision support system. During the 1970s, businesses also began to create routine reports to inform decision makers (managers) about what had happened in the previous period (e.g., day, week, month, quarter). Although it was useful to know what had happened in the past, managers needed more than this: They needed a variety of reports at different levels of granularity to better understand and address changing needs and challenges of the business.

The 1980s saw a significant change in the way organizations captured business-related data. The old practice had been to have multiple disjointed information systems tailored to capture transactional data of different organizational units or functions (e.g., accounting, marketing and sales, finance, manufacturing). In the 1980s, these systems were integrated as enterprise-level information systems that we now commonly call *ERP systems*. The old mostly sequential and nonstandardized data representation schemas were replaced by relational database management (RDBM) systems. These systems made it possible to improve the capture and storage of data, as well as the relationships between organizational data fields while significantly reducing the replication of information. The need for RDBM and ERP system emerged when data integrity and consistency became an issue, significantly hindering the effectiveness of business practices. With ERP, all the data from every corner of the enterprise is collected and integrated into a consistent schema so that every part of the organization has access to the single version of the truth when and where needed. In addition to the emergence of ERP systems—or perhaps because of these systems—business reporting became an on-demand, as-needed business practice. Decision makers could decide when they needed to or wanted to create specialized reports to investigate organizational problems and opportunities.

In the 1990s, the need for more versatile reporting led to the development of executive information systems (decision support systems designed and developed specifically for executives and their decision-making needs). These systems were designed as graphical dashboards and scorecards so that they could serve as visually appealing displays while focusing on the most important factors for decision makers to keep track of—the key performance indicators. In order to make this highly versatile reporting possible while keeping the transactional integrity of the business information systems intact, it was necessary to create a middle data tier—known as a *data warehouse* (*DW*)—as a repository to specifically support business reporting and decision making. In a very short time, most large to medium-size businesses adopted data warehousing as their platform for enterprise-wide decision making. The dashboards and scorecards got their data from a data warehouse, and by doing so, they were not hindering the efficiency of the business transaction systems—mostly referred to as enterprise resource planning (ERP) systems.

In the 2000s the DW-driven decision support systems began to be called *business intelligence systems*. As the amount of longitudinal data accumulated in the DWs increased, so did the capabilities of hardware and software to keep up with the rapidly changing and evolving needs of the decision makers. Because of the globalized competitive marketplace, decision makers needed current information in a very digestible format to address business problems and to take advantage of market opportunities in a timely manner. Because the data in a DW is updated periodically, it does not reflect the latest information. In order to elevate this information latency problem, DW vendors developed a system to update the data more frequently, which led to the terms *real-time data warehousing* and, more realistically, *right-time data warehousing*, which differs from the former by adopting a data refreshing policy based on the needed freshness of the data items (i.e., not all data items need to be refreshed in real time). Data warehouses are very large and feature rich, and it became necessary to "mine" the corporate data to "discover" new and useful knowledge nuggets to improve business processes and practices—hence the terms *data mining* and *text mining*. With the increasing volumes and varieties of data, the needs for more storage and more processing power emerged. While large corporations had the means to tackle this problem, small to medium-size companies needed financially more manageable business models. This need led to service-oriented architecture and software and infrastructure-as-a-service analytics business models. Smaller companies therefore gained access to analytics capabilities on an as-needed basis and paid only for what they used, as opposed to investing in financially prohibitive hardware and software resources.

In the 2010s we are seeing yet another paradigm shift in the way that data is captured and used. Largely because of the widespread use of the Internet, new data-generation mediums have emerged. Of all the new data sources (e.g., RFID tags, digital energy meters, clickstream Web logs, smart home devices, wearable health monitoring equipment), perhaps the most interesting and challenging is social networking/social media. This unstructured data is rich in information content, but analysis of such data sources poses significant challenges to computational systems, from both software and hardware perspectives. Recently, the term *Big Data* has been coined to highlight the challenges that these new data streams have brought upon us. Many advancements in both hardware (e.g., massively parallel processing with very large computational memory and highly parallel multiprocessor computing systems) and software/algorithms (e.g., Hadoop with MapReduce and NoSQL) have been developed to address the challenges of Big Data.

It's hard to predict what the next decade will bring and what the new analytics-related terms will be. The time between new paradigm shifts in information systems and particularly in analytics has been shrinking, and this trend will continue for the foreseeable future. Even though analytics is not new, the explosion in its popularity is very new. Thanks to the recent explosion in Big Data, ways to collect and store this data, and intuitive software tools, data and data-driven insight are more accessible to business professionals than ever before. Therefore, in the midst of global competition, there is a huge opportunity to make better managerial decisions by using data and analytics to increase revenue while decreasing costs by building better products, improving customer experience, and catching fraud before it happens, improving customer engagement through targeting and customization—all with the power of analytics and data. More and more companies are now preparing their employees with the know-how of business analytics to drive effectiveness and efficiency in their day-to-day decision-making processes.

A Simple Taxonomy for Analytics

Because of the multitude of factors related to both the need to make better and faster decisions and the availability and affordability of hardware and software technologies, analytics is gaining popularity faster than any other trends we have seen in recent history. Will this upward exponential trend continue? Many industry experts think it will, at least for the foreseeable future. Some of the most respected consulting companies are projecting that analytics will grow at three times the rate of other business segments in upcoming years; they have also named analytics as one of the top business trends of this decade (Robinson et al., 2010). As interest in and adoption of analytics have grown rapidly, a need to characterize analytics into a simple taxonomy has emerged. The top consulting companies (e.g., Accenture, Gartner, and IDT) and several technologically oriented academic institutions have embarked on a mission to create a simple taxonomy for analytics. Such a taxonomy, if developed properly and adopted universally, could create a contextual description of analytics, thereby facilitating a common understanding of what analytics is, including what is included in analytics and how analytics-related terms (e.g., business intelligence, predictive modeling, data mining) relate to each other. One of the academic institutions involved in this challenge is INFORMS (Institute for Operations Research and Management Science). In order to reach a wide audience, INFORMS hired Capgemini, a strategic management consulting firm, to carry out a study and characterize analytics.

The Capgemini study produced a concise definition of analytics: "Analytics facilitates realization of business objectives through reporting of data to analyze trends, creating predictive models for forecasting and optimizing business processes for enhanced performance." As this definition implies, one of the key findings from the study is that executives see analytics as a core function of businesses that use it. It spans many departments and functions within organizations, and in mature organizations, it spans the entire business. The study identified three hierarchical but sometimes overlapping groupings for analytics categories: descriptive, predictive, and prescriptive analytics. These three groups are hierarchical in terms of the level of analytics maturity of the organization. Most organizations start with descriptive analytics, then move into predictive analytics, and finally reach prescriptive analytics, the top level in the analytics hierarchy. Even though these three groupings of analytics are hierarchical in complexity and sophistication, moving from a lower level to a higher level is not clearly separable. That is, a business can be in the descriptive analytics level while at the same time using predictive and even prescriptive analytics capabilities, in a somewhat piecemeal fashion. Therefore, moving from one level to the next essentially means that the maturity at one level is completed and the next level is being widely exploited. Figure 1.3 shows a graphical depiction of the simple taxonomy developed by INFORMS and widely adopted by most industry leaders as well as academic institutions.

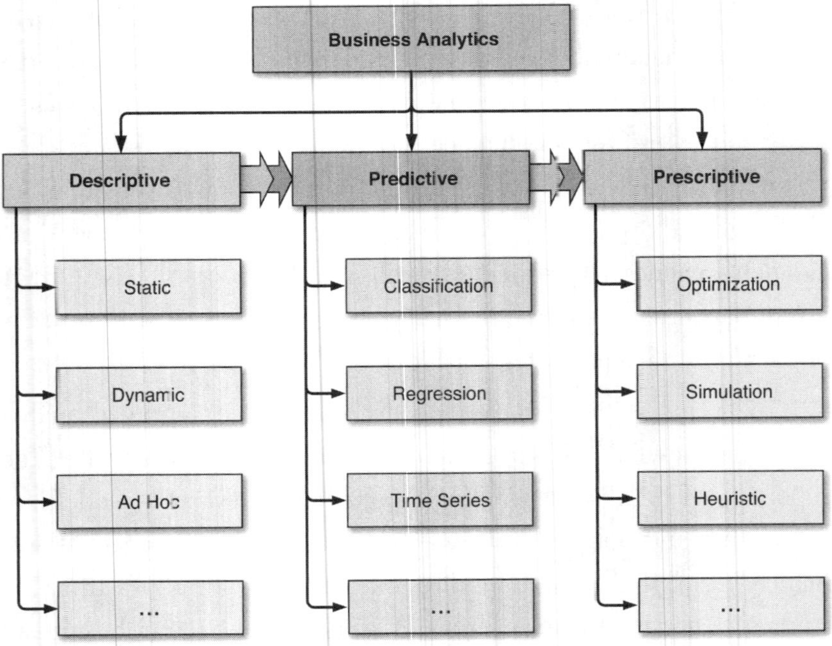

Figure 1.3 A Simple Taxonomy for Analytics

Descriptive analytics is the entry level in analytics taxonomy. It is often called *business reporting* because of the fact that most of the analytics activities at this level deal with creating reports to summarize business activities in order to answer questions such as "What happened?" and "What is happening?" The spectrum of these reports includes *static* snapshots of business transactions delivered to knowledge workers (i.e., decision makers) on a fixed schedule (e.g., daily, weekly, quarterly); *dynamic* views of business performance indicators delivered to managers and executives in a easily digestible form—often in a dashboard-looking graphical interface—on a continuous manner; and *ad hoc* reporting where the decision maker is given the capability of creating his or her own specific report (using an intuitive drag-and-drop graphical user interface) to address a specific or unique decision situation.

Descriptive analytics is also called *business intelligence* (*BI*), and predictive and prescriptive analytics are collectively called *advanced analytics*. The logic here is that moving from descriptive to predictive and/or prescriptive analytics is a significant shift in the level of sophistication and therefore warrants the label *advanced*. BI has been one of the most popular technology trends for information systems designed to support managerial decision making since the start of the century. It was popular (to some extent, it still is in some business circles) until the arrival of the analytics wave. BI is the entrance to the world of analytics, setting the stage and paving the way toward more sophisticated decision analysis. Descriptive analytics systems usually work off a data warehouse, which is a large database specifically designed and developed to support BI functions and tools.

Predictive analytics comes right after descriptive analytics in the three-level analytics hierarchy. Organizations that are mature in descriptive analytics move to this level, where they look beyond what happened and try to answer the question "What will happen?" In the following chapters, we will cover the predictive capabilities of these analytics techniques in depth as part of data mining; here we provide only a very short description of the main predictive analytics classes. Prediction essentially is the process of making intelligent/scientific estimates about the future values of some variables, like customer demand, interest rates, stock market movements, etc. If what is being predicted is a categorical variable, the act of prediction is called *classification*; otherwise, it is called *regression*. If the predicted variable is time dependent, the prediction process is often called *time-series forecasting*.

Prescriptive analytics is the highest echelon in analytics hierarchy. It is where the best alternative among many courses of action—that are usually created/identified by predictive and/or descriptive analytics—is determined using sophisticated mathematical models. Therefore, in a sense, this type of analytics tries to answer the question "What should I do?" Prescriptive analytics uses optimization-, simulation-, and heuristics-based decision-modeling techniques. Even though prescriptive analytics is at the top of the analytics hierarchy, the methods behind it are not new. Most of the optimization and simulation models that constitute prescriptive analytics were developed during and right after World War II, when there was a dire need for a lot with limited resources. Since then, some businesses have used these models for some very specific problem types, including yield/revenue management, transportation modeling, scheduling, etc. The new taxonomy of analytics has made them popular again, opening their use to a wide array of business problems and situations.

Figure 1.4 shows a tabular representation of the three hierarchical levels of analytics, along with the questions answered and techniques used at each level. As can be seen data mining is the key enabler of predictive analytics.

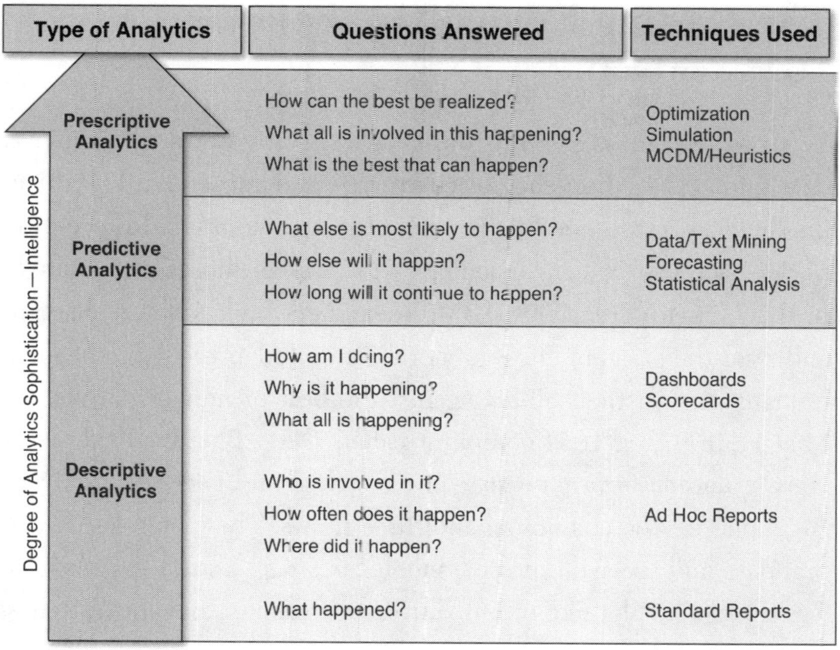

Figure 1.4 Three Levels of Analytics and Their Enabling Techniques

Business analytics is gaining popularity because it promises to provide decision makers with information and knowledge that they need to succeed. Effectiveness of business analytics systems, no matter the level in the analytics hierarchy, depends largely on the quality and quantity of the data (volume and representational richness); the accuracy, integrity, and timeliness of the data management system; and the capabilities and sophistication of the analytical tools and procedures used in the process. Understanding the analytics taxonomy helps organizations to be smart about selecting and implementing analytics capabilities to efficiently navigate through the maturity continuum.

The Cutting Edge of Analytics: IBM Watson

IBM Watson is perhaps the smartest computer system built to date. Since the emergence of computers and subsequently artificial intelligence in the late 1940s, scientists have compared the performance of these "smart" machines with human minds. Accordingly, in the mid- to late 1990s, IBM researchers built a smart machine and used the game of chess (generally credited as the game of smart humans) to test their ability against the best of human players. On May 11, 1997, an IBM computer called Deep Blue beat the world chess grandmaster after a six-game match series: two wins for Deep Blue, one for the champion, and three draws. The match lasted several days and received massive media coverage around the world. It was the classic plot line of human versus machine. Beyond the chess contest, the intention of developing this kind of computer intelligence was to make computers able to handle the kinds of complex calculations needed to help discover new medical drugs, do the broad financial modeling needed to identify trends and do risk analysis, handle large database searches, and perform massive calculations needed in advanced fields of science.

After a couple decades, IBM researchers came up with another idea that was perhaps more challenging: a machine that could not only play *Jeopardy!* but beat the best of the best. Compared to chess, *Jeopardy!* is much more challenging. While chess is well structured and has very simple rules, and therefore is very good match for computer processing, *Jeopardy!* is neither simple nor structured. *Jeopardy!* is a game designed for human intelligence and creativity, and therefore a computer designed to play it needed to be a cognitive computing system that can work and think like a human. Making sense of imprecision inherent in human language was the key to success.

In 2010 an IBM research team developed Watson, an extraordinary computer system—a novel combination of advanced hardware and software—designed to answer questions posed in natural human language. The team built Watson as part of the DeepQA project and named it after IBM's first president, Thomas J. Watson. The team that built Watson was looking for a major research challenge: one that could rival the scientific and popular interest of Deep Blue and would also have clear relevance to IBM's business interests. The goal was to advance computational science by exploring new ways for computer technology to affect science, business, and society at large. Accordingly, IBM Research undertook a challenge to build Watson as a computer system that could compete at the human champion level in real time on the American TV quiz show *Jeopardy!* The team wanted to create a real-time automatic contestant on the show, capable of listening, understanding, and responding—not merely a laboratory exercise.

Competing Against the Best at Jeopardy!

In 2011, as a test of its abilities, Watson competed on the quiz show *Jeopardy!*, in the first-ever human-versus-machine matchup for the show. In a two-game, combined-point match (broadcast in three *Jeopardy!* episodes during February 14–16), Watson beat Brad Rutter, the biggest all-time money winner on *Jeopardy!*, and Ken Jennings, the record holder for the longest championship streak (75 days). In these episodes, Watson consistently outperformed its human opponents on the game's signaling device, but it had trouble responding to a few categories, notably those having short clues containing only a few words. Watson had access to 200 million pages of structured and unstructured content, consuming 4 terabytes of disk storage. During the game, Watson was not connected to the Internet.

Meeting the *Jeopardy!* challenge required advancing and incorporating a variety of text mining and natural language processing technologies, including parsing, question classification, question decomposition, automatic source acquisition and evaluation, entity and relationship detection, logical form generation, and knowledge representation and reasoning. Winning at *Jeopardy!* requires accurately computing confidence in answers. The questions and content are ambiguous and noisy, and none of the individual algorithms are perfect. Therefore, each component must produce a confidence in its output, and individual component confidences must be combined to compute the overall confidence of the final answer. The final confidence is used to determine whether the computer system should risk choosing to answer at all. In *Jeopardy!* parlance, this confidence is used to determine whether the computer will "ring in" or "buzz in" for a question. The confidence must be computed during the time the question is read and before the opportunity to buzz in. This is roughly between one and six seconds, with an average around three seconds.

How Does Watson Do It?

The system behind Watson, which is called DeepQA, is a massively parallel, text mining–focused, probabilistic evidence-based computational architecture. For the *Jeopardy!* challenge, Watson used more than 100 different techniques for analyzing natural language, identifying sources, finding and generating hypotheses, finding and scoring evidence, and merging and ranking hypotheses. What is far more important than any particular technique the IBM team used was how it combined them in DeepQA such that overlapping approaches could bring their strengths to bear and contribute to improvements in accuracy, confidence, and speed.

DeepQA is an architecture with an accompanying methodology that is not specific to the *Jeopardy!* challenge. These are the overarching principles in DeepQA:

- **Massive parallelism.** Watson needed to exploit massive parallelism in the consideration of multiple interpretations and hypotheses.

- **Many experts.** Watson needed to be able to integrate, apply, and contextually evaluate a wide range of loosely coupled probabilistic question and content analytics.
- **Pervasive confidence estimation.** No component of Watson commits to an answer; all components produce features and associated confidences, scoring different question and content interpretations. An underlying confidence-processing substrate learns how to stack and combine the scores.
- **Integration of shallow and deep knowledge.** Watson needed to balance the use of strict semantics and shallow semantics, leveraging many loosely formed ontologies.

Figure 1.5 illustrates the DeepQA architecture at a very high level. More technical details about the various architectural components and their specific roles and capabilities can be found in Ferrucci et al. (2010).

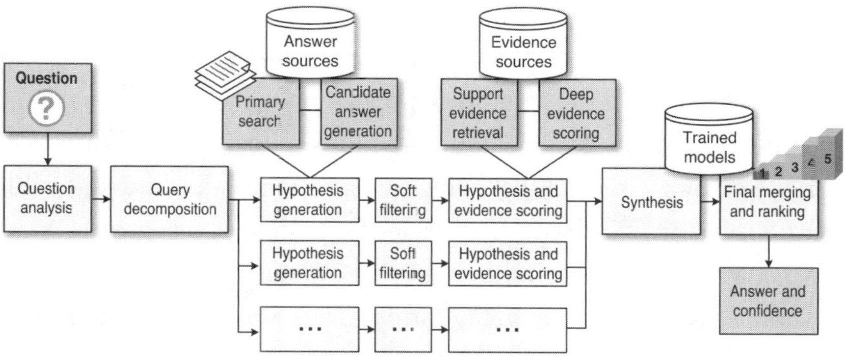

Figure 1.5 A High-Level Depiction of DeepQA Architecture

What Is the Future for Watson?

The *Jeopardy!* challenge helped IBM address requirements that led to the design of the DeepQA architecture and the implementation of Watson. After three years of intense research and development by a core team of about 20 researchers, as well as a significant R&D budget, Watson managed to perform at human expert levels in terms of precision, confidence, and speed at the *Jeopardy!* quiz show.

After the show, the big question was "So what now?" Was developing Watson all for a quiz show? Absolutely not! Showing the rest of the world what Watson (and the cognitive system behind it) could do became an inspiration for the next generation of intelligent information systems. For IBM, it was a demonstration of what is possible with cutting-edge analytics and computational sciences. The message is clear: If a smart machine can beat the best of the best in humans at what they are the best at, think about what it can do for your organizational problems. The first industry that utilized Watson was health care, followed by security, finance, retail, education, public services, and research. The following sections provide short descriptions of what Watson can do (and, in many cases, is doing) for these industries.

Health Care

The challenges that health care is facing today are rather big and multifaceted. With the aging U.S. population, which may be partially attributed to better living conditions and advanced medical discoveries fueled by a variety of technological innovations, demand for health care services is increasing faster than the supply of resources. As we all know, when there is an imbalance between demand and supply, the prices go up and the quality suffers. Therefore, we need cognitive systems like Watson to help decision makers optimize the use of their resources, both in clinical and managerial settings.

According to health care experts, only 20% of the knowledge physicians use to diagnose and treat patients is evidence based. Considering that the amount of medical information available is doubling every five years and that much of this data is unstructured, physicians simply don't have time to read every journal that can help them keep up-to-date with the latest advances. Given the growing demand for services and the complexity of medical decision making, how can health care providers address these problems? The answer could be to use Watson, or some other cognitive systems like Watson that has the ability to help physicians in diagnosing and treating patients by analyzing large amounts of data—both structured data coming from electronic medical record databases and unstructured text coming from physician notes and published literature—to provide evidence for faster and better decision making. First, the physician and the patient can describe symptoms and other related factors to the system in natural language. Watson can then identify the key pieces of information and mine the patient's data to find relevant facts about family history, current medications, and other existing conditions. It can then combine that information with current findings from tests, and then it can form and test hypotheses for potential diagnoses by examining a variety of data sources—treatment guidelines, electronic medical record data and doctors' and nurses' notes, and peer-reviewed research and clinical studies. Next, Watson can suggest potential diagnostics and treatment options, with a confidence rating for each suggestion.

Watson also has the potential to transform health care by intelligently synthesizing fragmented research findings published in a variety of outlets. It can dramatically change the way medical students learn. It can help healthcare managers to be proactive about the upcoming demand patterns, optimally allocate resources, and improve processing of payments. Early examples of leading health care providers that use Watson-like cognitive systems include MD Anderson, Cleveland Clinic, and Memorial Sloan Kettering.

Security

As the Internet expands into every facet of our lives—ecommerce, ebusiness, smart grids for energy, smart homes for remote control of residential gadgets and appliances—to make things easier to manage, it also opens up the potential for ill-intended people to intrude in our lives. We need smart systems like Watson that are capable of constantly monitoring for abnormal behavior and, when it is identified, preventing people from accessing our lives and harming us. This could be at the corporate or even national security system level; it could also be at the personal level. Such a smart system could learn who we are and become a digital guardian that could make inferences about activities related to our life and alert us whenever abnormal things happen.

Finance

The financial services industry faces complex challenges. Regulatory measures, as well as social and governmental pressures for financial institutions to be more inclusive, have increased. And the customers the industry serves are more empowered, demanding, and sophisticated than ever before. With so much financial information generated each day, it is difficult to properly harness the right information to act on. Perhaps the solution is to create smarter client engagement by better understanding risk profiles and the operating environment. Major financial institutions are already working with Watson to infuse intelligence into their business processes. Watson is tackling data-intensive challenges across the financial services sector, including banking, financial planning, and investing.

Retail

Retail industry is rapidly changing with the changing needs and wants of customers. Customers, empowered by mobile devices and social networks that give them easier access to more information faster than ever before, have high expectations for products and services. While retailers are using analytics to keep up with those expectations, their bigger challenge is efficiently and effectively analyzing the growing mountain of real-time insights that could give them the competitive advantage. Watson's cognitive computing capabilities related to analyzing massive amounts of unstructured data can help retailers reinvent their decision-making processes around pricing, purchasing, distribution, and staffing. Because of Watson's ability to understand and answer questions in natural language, it is an effective and scalable solution for analyzing and responding to social sentiment based on data obtained from social interactions, blogs, and customer reviews.

Education

With the rapidly changing characteristics of students—more visually oriented/stimulated, constantly connected to social media and social networks, increasingly shorter attention spans—what should the future of education and the classroom look like? The next generation of educational system should be tailored to fit the needs of the new generation, with customized learning plans, personalized textbooks (digital ones with integrated multimedia—audio, video, animated graphs/charts, etc.), dynamically adjusted curriculum, and perhaps smart digital tutors and 24/7 personal advisors. Watson seems to have what it takes to make all this happen. With its natural language processing capability, students can converse with it just as they do with their teachers, advisors, and friends. This smart assistant can answer students' questions, satisfy their curiosity, and help them keep up with the endeavors of the educational journey.

Government

For local, regional, and national governments, the exponential rise of Big Data presents an enormous dilemma. Today's citizens are more informed and empowered than ever before, and that means they have high expectations for the value of the public sector serving them. And government organizations can now gather enormous volumes of unstructured, unverified data that could serve their citizens—but only if that data can be analyzed efficiently and effectively. IBM Watson's cognitive computing may help make sense of this data deluge, speeding governments' decision-making processes and helping public employees focus on innovation and discovery.

Research

Every year, hundreds of billions of dollars are spent on research and development, most of it documented in patents and publications, creating enormous amount of unstructured data. To contribute to the extant knowledgebase, one needs to sift through these data sources to find the outer boundaries of research in a particular field. This is very difficult, if not impossible work, if it is done with traditional means, but Watson can act as a research assistant to help collect and synthesize information to keep people updated on recent findings and insights. For instance, New York Genome Center is using the IBM Watson cognitive computing system to analyze the genomic data of patients diagnosed with a highly aggressive and malignant brain cancer and to more rapidly deliver personalized, life-saving treatment to patients with this disease (Royyuru, 2014).

References

Bi, R. (2014). *When Watson Meets Machine Learning*. www.kdnuggets.com/2014/07/watson-meets-machine-learning.html (accessed June 2014).

DeepQA. (2011). *DeepQA Project: FAQ, IBM Corporation*. www.research.ibm.com/deepqa/faq.shtml (accessed April 2014).

Feldman, S., J. Hanover, C. Burghard, & D. Schubmehl. (2012). *Unlocking the Power of Unstructured Data*. www-01.ibm.com/software/ebusiness/jstart/downloads/unlockingUnstructuredData.pdf (accessed May 2014).

Ferrucci, D., et al. (2010). "Building Watson: An Overview of the DeepQA Project," *AI Magazine*, 31(3).

IBM. (2014). *Implement Watson*. www.ibm.com/smarterplanet/us/en/ibmwatson/implement-watson.html (accessed July 2014).

Liberatore, M., & W. Luo. (2011). "INFORMS and the Analytics Movement: The View of the Membership," *Interfaces*, 41(6): 578–589.

Robinson, A., J. Levis, & G. Bennett. (2010, October). "Informs to Officially Join Analytics Movement," *ORMS Today*.

Royyuru, A. (2014). "IBM's Watson Takes on Brain Cancer: Analyzing Genomes to Accelerate and Help Clinicians Personalize Treatments." Thomas J. Watson Research Center, www.research.ibm.com/articles/genomics.shtml (accessed September 2014).

2

Introduction to Data Mining

Data mining is one of the most powerful enablers of analytics. Although its roots date back to the late 1980s and early 1990s, the most impactful data mining applications have been developed after the turn of the 21st century. Many people believe that the recent popularity of analytics can largely be credited to the increasing use of data mining, which is extracting and providing much-needed insight and knowledge to decision makers at all levels of the managerial hierarchy. The term *data mining* was originally used to describe the process through which previously unknown patterns in data were discovered. Software vendors and consulting companies have expanded this definition to include most forms of data analysis in order to increase its reach and capability and thereby increase sales of software tools and services related to data mining. With the emergence of *analytics* as an overarching term for all data analyses, data mining is put back into its proper place—on the analytics continuum where the new discovery of knowledge happens.

In an article in *Harvard Business Review*, Thomas Davenport (2006), a well-known and respected expert in the field of analytics, argued that the latest strategic weapon for today's businesses is analytical decision making that is based on discovery of new knowledge through data mining. He provided examples of companies such as Amazon.com, Capital One, Marriott International, and others that have used (and are still using) analytics to better understand their customers and optimize their extended supply chains in order to maximize their returns on investment while providing the best possible customer service. This level of success can happen only if the company pursues all avenues, including analytics at all three levels—descriptive, predictive, and prescriptive—to intimately understand its customers and their needs and wants, along with their vendors, business processes, and the extended supply chain.

Data mining is the process of converting data into information and to knowledge. In the context of knowledge management, data mining is the phase in which new knowledge is created. Comparatively speaking, knowledge is very distinct from data and information (see Figure 2.1). Whereas data is facts, measurements, and statistics, information is organized or processed data that is timely (i.e., inferences from the data are drawn within the time frame of applicability) and understandable (i.e., with regard to the original data). Knowledge is information that is contextual, relevant, and actionable. For example, a map that gives detailed driving directions from one location to another could be considered data. An up-to-the-minute traffic bulletin along the freeway that indicates a traffic slowdown due to construction several miles ahead could be considered information. Awareness of an alternative, back-roads route could be considered knowledge. In this case, the map is considered data because it does not contain current relevant information that affects the driving time and conditions from one location to the other. However, having the current conditions as information is useful only if you have knowledge that enables you to avoid the construction zone. The implication is that knowledge has strong experiential and reflective elements that distinguish it from information in a given context.

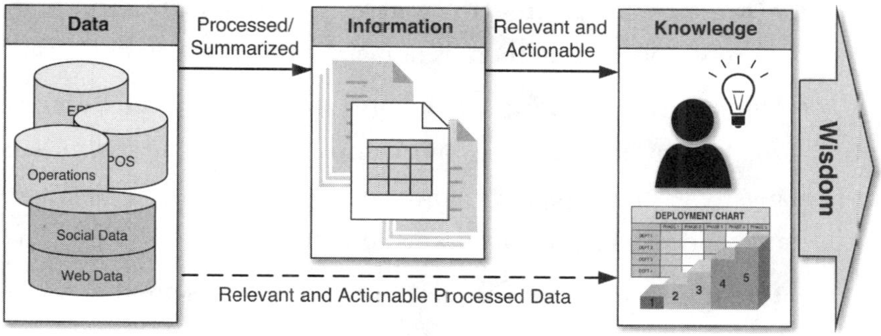

Figure 2.1 The Continuum of Data to Information to Knowledge

Having knowledge implies that it can be exercised to solve a problem, whereas having information does not carry the same connotation. An ability to act is an integral part of being knowledgeable. For example, two people in the same context with the same information may not have the same ability to use the information with the same degree of success. Hence, there is a difference in the human capability to add value. The differences in ability may be due to different experiences, different training, different perspectives, and other factors. Whereas data, information, and knowledge may all be viewed as assets of an organization, knowledge provides a higher level of meaning about data and information. Because it conveys meaning, it tends to be much more valuable yet more ephemeral.

Although the term *data mining* is relatively new to many people, the ideas behind it are not. Many of the techniques used in data mining have their roots in traditional statistical analysis and artificial intelligence work done since the early part of 1950s. Why, then, has data mining suddenly gained the attention of the business world? Following are some of the most common reasons:

- **More intense competition at the global scale.** There are more suppliers than there is demand to satisfy everybody.
- **Constantly changing needs and wants of the customers.** Because of the increasing number of suppliers and their offerings (e.g., higher quality, lower cost, faster service), customer demand is changing in a dramatic fashion.
- **Recognition of the value of data.** Businesses are now aware of the untapped value hidden in large data sources.
- **Changing culture of management.** Data-driven, evidence-based decision making is becoming a common practice, significantly changing the way managers work.
- **Improved data capture and storage techniques.** Collection and integration of data from a variety of sources into standardized data structures enable businesses to have quality data about customers, vendors, and business transactions.
- **Emergence of data warehousing.** Databases and other data repositories are being consolidated into a single location in the form of a data warehouse to support analytics and managerial decision making.
- **Technological advancements in hardware and software.** Processing and storage capabilities of computing devices are increasing exponentially.
- **Cost of ownership.** While capabilities are increasing, the cost of hardware and software for data storage and processing are rapidly decreasing.
- **Availability of data.** Living in the age of Internet brings new opportunities to analytically savvy businesses to identify and tap into very large and information-rich data sources (social networks and social media) to better understand the world they are living in.

Data is everywhere. For instance, data generated by Internet-based activities is increasing rapidly, reaching volumes that we did not even have specific names for in the very recent past. Large amounts of genomic data and related information (in the form of publications and research findings published in journal articles and other media outlets) are being generated and accumulated all over the world. Disciplines such as astronomy and nuclear physics create huge quantities of data on a regular basis. Medical and pharmaceutical researchers constantly generate and store data that can then be used in data mining applications to identify better ways to diagnose and treat illnesses and to discover new and improved drugs. On the commercial side, perhaps the most common use of data and data mining has been in the finance, retail, and health care sectors. Data mining is used to detect and reduce fraudulent activities, especially in insurance claims and credit card use; to identify customer buying patterns; to reclaim profitable customers; to identify trading rules from historical data; and to aid in increased profitability using market-basket analysis. Data mining is already widely used to better target clients, and with the widespread development of ecommerce, this can only become more imperative with time.

What Is Data Mining?

At the most basic level, *data mining* can be defined as the process of discovering (i.e., mining) knowledge (i.e., actionable information) from large amounts of data. When you really think about it, you realize that the term *data mining* is not a proper description of what's really happening; that is, mining of gold from within rocks and dirt is referred to as *gold* mining rather than *rock* or *dirt* mining. Therefore, data mining perhaps should have been named *knowledge mining* or *knowledge discovery*. Despite the mismatch between the term and its meaning, data mining has become the term of choice in the community at large. Several other names, including *knowledge discovery in databases*, *information extraction*, *pattern analysis*, *information harvesting*, and *pattern searching*, among others, have been suggested to replace data mining, but none has gained any significant traction.

Data mining is a process that uses statistical, mathematical, and artificial intelligence techniques and algorithms to extract and identify useful information and subsequent knowledge (or patterns) from large sets of data. These patterns can be in the form of business rules, affinities, correlations, trends, or predictions. Fayyad et al. (1996) define data mining as "the nontrivial process of identifying valid, novel, potentially useful, and ultimately understandable patterns in data stored in structured databases," where the data is organized in records structured by categorical, ordinal, and continuous variables. The key terms in this definition have the following meanings:

- *Process* implies that data mining comprises many iterative steps.
- *Nontrivial* means that some experimentation-type search or inference is involved; that is, it is not as straightforward as a computation of predefined quantities.
- *Valid* means that the discovered patterns should hold true on new data with a sufficient degree of certainty.
- *Novel* means that the patterns were not previously known to the user in the context of the system being analyzed.
- *Potentially useful* means that the discovered patterns should lead to some benefit to the user or task.
- *Ultimately understandable* means that the pattern should make business sense that leads to users saying "This makes sense. Why didn't I think of that?"—if not immediately, at least after some processing.

Data mining is not a new discipline but rather a new approach in the intersection of several other scientific disciplines. To some extent, data mining is a new philosophy that suggests the use of data and mathematical models to create/discover new knowledge. Data mining leverages capabilities of other disciplines, including statistics, artificial intelligence, machine learning, management science, information systems, and databases, in a systematic and synergistic way (see Figure 2.2). Using collective capabilities of these scientific disciplines, data mining aims to make progress in extracting useful information and knowledge from large data repositories. It is an emerging field that has attracted much attention in a very short time and has fueled the emergence and popularity of the analytics movement.

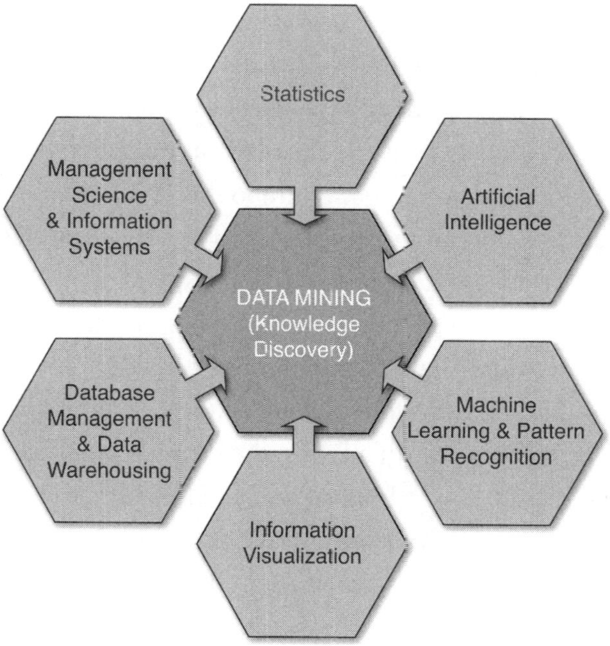

Figure 2.2 Data Mining: A Multidisciplinary Approach to Knowledge Discovery

What Data Mining Is Not

Because of its appeal, many have used the term *data mining* to mean any analysis that has something to do with data. For instance, people may refer to casual Internet searching as data mining. It's true that Internet searching is a way to mine large and diverse data/information sources to find (i.e., discover) evidence for a specific question/query; and because of that, it may seem like data mining. However, data mining is the process of discovering reusable patterns through the application of statistical and/or machine learning techniques. Therefore, data mining is much more rigorous and scientific than simply querying the Internet.

Another concept that data mining is often confused with is OLAP (online analytical processing). As the core enabler of the business intelligence movement, OLAP is a collection of database-querying methods for searching very large databases (or data warehouses) through the use of data cubes. Cubes are multidimensional representations of the data stored in data warehouses. With the use of cubes, OLAP helps decision makers to slice and dice organizational data for the purpose of finding answers to "What happened?" "Where did it happen?" and "When did it happen?" types of questions. As sophisticated as it may sound—and perhaps from the standpoint of efficiency, it actually is that sophisticated—OLAP is not data mining. It may be a precursor to data mining. In a sense, they complement each other in that they convert data into information and knowledge for better and faster decision making. Whereas OLAP is part of descriptive analytics, data mining is an essential part of predictive analytics.

There has been much discussion about statistics and data mining; some people suggest that data mining is a part of statistics and others propose that statistics is a part of data mining. Yet others suggest that data mining and statistics are the same thing. Even though we cannot get to the bottom of that discussion here, we can at least shed some light to it by mentioning a few critical points. Data mining and statistics have a lot in common. They both look for relationships in data. The main difference between the two is that statistics starts with a well-defined proposition and hypothesis, while data mining starts with a loosely defined discovery statement. Statistics collects a sample of data (i.e., primary data) to test the hypothesis, while data mining and analytics use existing data (i.e., often observational, secondary data) to discover novel patterns and relationships. Another difference is in the size of data that they use. Data mining looks for data sets that are as "big" as possible, while statistics looks for right size of data and, if the data is larger than what is needed or required for the statistical analysis, it uses a sample of the data. Statistics and data mining have deferent definitions of what constitutes "large data": Whereas a few hundred to a thousand data points is large enough to a statistician, several millions to a few billion data points is considered large for data mining studies.

In summary, data mining is not a simple Internet search or routine application of OLAP, and it's not the same as statistics. Even though it uses capabilities of these descriptive techniques, data mining is the next level in the analytics hierarchy, where interesting patterns (relationships and future trends) are discovered with the use of data and models.

The Most Common Data Mining Applications

Data mining has become a popular tool in addressing many complex business problems and opportunities. It has proven to be very successful and helpful in many areas, some of which are listed and briefly discussed in the following sections. It is difficult to find an industry or a problem area where a significant amount of data mining applications have not already been covered in the literature. The goal of many of these data mining applications is to solve complex problems or to explore emerging opportunities in order to create a sustainable competitive advantage.

Marketing and Customer Relationship Management

Customer relationship management (CRM) is an extension of traditional marketing. The goal of CRM is to create one-on-one relationships with customers by developing an intimate understanding of their needs and wants. As businesses build relationships with their customers over time through a variety of interactions (e.g., product inquiries, sales, service requests, warranty calls, product reviews, social media connections), they accumulate tremendous amounts of data. When combined with demographic and socioeconomic attributes, this information-rich data can be used to identify most likely responders/buyers of new products and services (i.e., customer profiling); understand the root causes of customer attrition in order to improve customer retention (i.e., churn analysis); discover time-variant associations between products and services to maximize sales and customer value; and find the most profitable customers and their preferential needs to strengthen relationships and maximize sales.

Banking and Finance

Data mining can help banks and other financial institutions address a variety of problems and opportunities. Data mining can be used to streamline and automate the processing of loan applications by accurately predicting and identifying the most probable defaulters; to detect fraudulent credit card and online-banking transactions; to find new ways to maximize customer value by selling products and services that customers are most likely to buy; and to optimize cash return by accurately forecasting the cash flow on banking entities (e.g., ATMs, bank branches).

Retailing and Logistics

In retailing, data mining can be used to predict accurate sales volumes at specific retail locations in order to determine correct inventory levels; to identify sales relationships between different products (with market-basket analysis) to improve store layout and optimize sales promotions; to forecast consumption levels of different product types (based on seasonal and environmental conditions); to optimize logistics and hence maximize sales; and to discover interesting patterns in the movement of products (especially for products that have a limited shelf life because they are prone to expiration, perishability, and contamination) in a supply chain by analyzing sensory and RFID data.

Manufacturing

Manufacturers can use data mining to predict machinery failures before they occur through the use of sensory data (enabling *condition-based maintenance*); to identify anomalies and commonalities in production systems to optimize manufacturing capacity; and to discover novel patterns to identify and improve product quality.

Brokerages and Securities Trading

Brokers and traders use data mining to predict when and how much certain stock and/or bond prices will change; to forecast the range and direction of market fluctuations; to assess the effect of particular issues and events on overall market movements; and to identify and prevent fraudulent activities in securities trading.

Insurance

The insurance industry uses data mining techniques to forecast claim amounts for property and medical coverage costs for better business planning; to determine optimal rate plans based on the analysis of claims and customer data; to predict which customers are most likely to buy new policies with special features; and to identify and prevent incorrect claims payments and fraudulent activities.

Computer Hardware and Software

Data mining can be used to predict disk drive failures well before they actually occur; to identify and filter unwanted Web content and email messages; to detect and prevent computer network security bridges; and to identify potentially unsecure software products.

Government and Defense

Data mining has a number of government and military applications. It can be used to forecast the cost of moving military personnel and equipment; to predict an adversary's moves and hence develop more successful strategies for military engagements; to predict resource consumption for better planning and budgeting; and to identify classes of unique experiences, strategies, and lessons learned from military operations for better knowledge sharing throughout the organization.

Travel and Lodging

Data mining has a variety of uses in the travel industry. It is successfully used to predict sales of different services (e.g., seat types in airplanes, room types in hotels/resorts, car types in rental car companies) in order to optimally price services to maximize revenues as a function of time-varying transactions (commonly referred to as *yield management*); to forecast demand at different locations to better allocate limited organizational resources; to identify the most profitable customers and provide them with personalized services to maintain their repeat business; and to retain valuable employees by identifying and acting on the root causes for attrition.

Health and Health Care

Data mining has a number of health care applications. It can be used to help individuals and groups pursue healthier life styles (by analyzing data collected with wearable health-monitoring devices); to identify people without health insurance and the factors underlying this undesired phenomenon; to identify novel cost–benefit relationships between different treatments to develop more effective strategies; to forecast the level and the time of demand at different service locations to optimally allocate organizational resources; and to understand the underlying reasons for customer and employee attrition.

Medicine

The use of data mining in medicine is an invaluable complement to traditional medical research, which is mainly clinical and biological in nature. Data mining analyses can be used to identify novel patterns to improve survivability of patients with cancer; to predict success rates of organ transplantation patients to develop better donor–organ matching policies; to identify the functions of different genes in the human chromosome (known as *genomics*); and to discover the relationships between symptoms and illnesses (as well as illnesses and successful treatments) to help medical professionals make informed decisions in a timely manner.

Entertainment Industry

Data mining is successfully used in the entertainment industry to analyze viewer data to decide what programs to show during prime time and how to maximize returns by knowing where to insert advertisements; to predict the financial success of movies before they are produced to make investment decisions and to optimize returns; to forecast the demand at different locations and different times to better schedule entertainment events and to optimally allocate resources; and to develop optimal pricing policies to maximize revenues.

Homeland Security and Law Enforcement

Data mining has a number of homeland security and law enforcement applications. Data mining is often used to identify patterns of terrorist behaviors; to discover crime patterns (e.g., locations, timings, criminal behaviors, and other related attributes) to help solve criminal cases in a timely manner; to predict and eliminate potential biological and chemical attacks on the nation's critical infrastructure by analyzing special-purpose sensory data; and to identify and stop malicious attacks on critical information infrastructures (often called *information warfare*).

Sports

Data mining was used to improve the performance of National Basketball Association (NBA) teams in the United States. Major League Baseball teams use predictive analytics and data mining to optimally utilize their limited resources for a winning season. (In fact, *Moneyball* is a popular movie about the use of analytics in baseball.) Most professional sports today employ data crunchers and use data mining to increase their chances of winning. Data mining applications are not limited to professional sports. In a recently published article, Delen at al. (2012) developed models to predict NCAA Bowl Game outcomes, using a wide range of variables about the two opposing teams' previous games statistics. Wright (2012) used a variety of predictors for examination of the NCAA men's basketball championship bracket (a.k.a. March Madness). In short, data mining can be used to predict the outcomes of sporting events, to identify means to increase odds of winning against a specific opponent, and to make the most out of the available resources (e.g., financial, managerial, athletic) so that a team can produce the best possible outcomes.

What Kinds of Patterns Can Data Mining Discover?

Using the most relevant data (which may come from organizational databases or may be obtained from outside sources), data mining builds models to identify patterns among the attributes (i.e., variables or characteristics) that exist in a data set. Models are usually the mathematical representations (simple linear correlations and/or complex highly nonlinear relationships) that identify the relationships among the attributes of the objects (e.g., customers) described in the data set. Some of these patterns are explanatory (explaining the interrelationships and affinities among the attributes), whereas others are predictive (projecting future values of certain attributes). In general, data mining seeks to identify three major types of patterns:

- *Associations* find commonly co-occurring groupings of things, such as "beers and diapers" or "bread and butter" commonly purchased and observed together in a shopping cart (i.e., market-basket analysis). Another type of association pattern captures sequences of things. These sequential relationships can discover time-ordered events, such as predicting that an existing banking customer who already has a checking account will open a savings account followed by an investment account within a year.
- *Predictions* tell the nature of future occurrences of certain events based on what has happened in the past, such as predicting the winner of the Super Bowl or forecasting the absolute temperature on a particular day.

- *Clusters* identify natural groupings of things based on their known characteristics, such as assigning customers in different segments based on their demographics and past purchase behaviors.

These types of patterns have been manually extracted from data by humans for centuries, but the increasing volume of data in modern times has created a need for more automatic approaches. As data sets have grown in size and complexity, direct manual data analysis has increasingly been augmented with indirect, automatic data processing tools that use sophisticated methodologies, methods, and algorithms. The manifestation of such evolution of automated and semi-automated means of processing large data sets is now commonly referred to as *data mining*.

As mentioned earlier, generally speaking, data mining tasks and patterns can be classified into three main categories: prediction, association, and clustering. Based on the way in which the patterns are extracted from the historical data, the learning algorithms of data mining methods can be classified as either supervised or unsupervised. With supervised learning algorithms, the training data includes both the descriptive attributes (i.e., independent variables or decision variables) as well as the class attribute (i.e., output variable or result variable). In contrast, with unsupervised learning, the training data includes only the descriptive attributes. Figure 2.3 shows a simple taxonomy for data mining tasks, along with the learning methods and popular algorithms for each of the data mining tasks. Out of the three main categories of tasks, prediction patterns/models can be classified as the outcome of a supervised learning procedure, while association and clustering patterns/models can be classified as the outcome of unsupervised learning procedures.

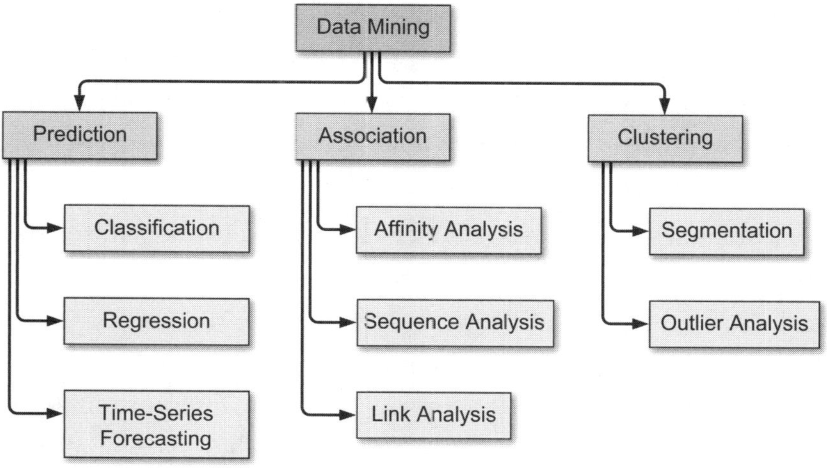

Figure 2.3 A Simple Taxonomy for Data Mining

Prediction is commonly used to indicate telling about the future. It differs from simple guessing by taking into account the experiences, opinions, and other relevant information in conducting the task of foretelling. A term that is commonly associated with prediction is *forecasting*. Even though many people use these two terms synonymously, there is a subtle difference between them. Whereas prediction is largely experience and opinion based, forecasting is data and model based. That is, in the order of increasing reliability, one might list the relevant terms as guessing, predicting, and forecasting. In data mining terminology, *prediction* and *forecasting* are used synonymously, and the term *prediction* is used as the common representation of the act. Depending on the nature of what is being predicted, prediction can be named more specifically as classification (where the predicted thing, such as tomorrow's forecast, is a class label such as "rainy" or "sunny") or regression (where the predicted thing, such as tomorrow's temperature, is a real number, such as "65°F").

Classification, or supervised induction, is perhaps the most common of all data mining tasks. The objective of classification is to analyze the historical data stored in a database and automatically generate a model that can predict future behavior. This induced model consists of generalizations over the records of a training data set, which help distinguish predefined classes. The hope is that the model can then be used to predict the classes of other unclassified records and, more importantly, to accurately predict actual future events.

Common classification tools include neural networks and decision trees (from machine learning), logistic regression and discriminant analysis (from traditional statistics), and emerging tools such as rough sets, support vector machines, and genetic algorithms. Statistics-based classification techniques (e.g., logistic regression, discriminant analysis) have received their share of criticism—that they make unrealistic assumptions about the data, such as independence and normality—which limit their use in classification-type data mining projects.

Neural networks (see Chapter 5, "Data Mining Algorithms," for more detailed coverage of this popular machine learning algorithm) involve the development of mathematical structures (somewhat resembling the biological neural networks in the human brain) that have the capability to learn from past experiences, presented in the form of well-structured data sets. They tend to be more effective when the number of variables involved is rather large and the relationships among them are complex and imprecise. Neural networks have disadvantages as well as advantages. For example, it is usually very difficult to provide good rationale for the predictions made by a neural network. Also, neural networks tend to need considerable training. Unfortunately, the time needed for training tends to increase exponentially as the volume of data increases, and, in general, neural networks cannot be trained on very large databases. These and other factors have limited the applicability of neural networks in data-rich domains.

Decision trees classify data into a finite number of classes, based on the values of the input variables. Decision trees are essentially a hierarchy of if–then statements and are thus significantly faster than neural networks. They are most appropriate for categorical and interval data. Therefore, incorporating continuous variables into a decision tree framework requires discretization—that is, converting continuous valued numerical variables to ranges and categories.

A related category of classification tools is rule induction. Unlike with a decision tree, with rule induction, the if–then statements are induced from the training data directly, and they need not be hierarchical in nature. Other, more recent techniques such as SVM, rough sets, and genetic algorithms are gradually finding their way into the arsenal of classification algorithms and are covered in more detail in Chapter 5 as part of the discussion on data mining algorithms.

Using *associations*, more commonly called *association rules* in data mining, is a popular and well-researched technique for discovering interesting relationships among variables in large databases. Thanks to automated data-gathering technologies such as bar code scanners, the use of association rules for discovering regularities among products in large-scale transactions recorded by point-of-sale systems in supermarkets has become a common knowledge-discovery task in the retail industry. In the context of the retail industry, association rule mining is often called *market-basket analysis*.

Two commonly used derivatives of association rule mining are link analysis and sequence mining. With link analysis, the linkage among many objects of interest is discovered automatically, such as the link between Web pages and referential relationships among groups of academic publication authors. With sequence mining, relationships are examined in terms of their order of occurrence to identify associations over time. Algorithms used in association rule mining include the popular Apriori (where frequent item sets are identified), FP-Growth, OneR, ZeroR, and Eclat. Chapter 4, "Data and Methods in Data Mining," provides an explanation of Apriori.

Clustering partitions a collection of things (e.g., objects, events, etc., presented in a structured data set) into segments (or natural groupings) whose members share similar characteristics. Unlike in classification, in clustering, the class labels are unknown. As the selected algorithm goes through the data set, identifying the commonalities of things based on their characteristics, the clusters are established. Because the clusters are determined using a heuristic-type algorithm, and because different algorithms may end up with different sets of clusters for the same data set, before the results of clustering techniques are put to actual use, it may be necessary for an expert to interpret and potentially modify the suggested clusters. After reasonable clusters have been identified, they can be used to classify and interpret new data.

Not surprisingly, clustering techniques include optimization. The goal of clustering is to create groups so that the members within each group have maximum similarity and the members across groups have minimum similarity. The most commonly used clustering techniques include *k*-means (from statistics) and self-organizing maps (from machine learning), which is a unique neural network architecture developed by Kohonen (1982).

Firms often effectively use their data mining systems to perform market segmentation with cluster analysis. Cluster analysis is a means of identifying classes of items so that items in a cluster have more in common with each other than with items in other clusters. This type of analysis can be used in segmenting customers and directing appropriate marketing products to the segments at the right time in the right format at the right price. Cluster analysis is also used to identify natural groupings of events or objects so that a common set of characteristics of these groups can be identified to describe them.

Two techniques often associated with data mining are visualization and time-series forecasting. Visualization can be used in conjunction with other data mining techniques to gain a clearer understanding of underlying relationships. As the importance of visualization has increased in recent years, the term *visual analytics* has emerged. The idea is to combine analytics and visualization in a single environment for easier and faster knowledge creation. Visual analytics is covered in detail in Chapter 4. In time-series forecasting, the data consists of values of the same variable that is captured and stored over time, in regular intervals. This data is then used to develop forecasting models to extrapolate the future values of the same variable.

Popular Data Mining Tools

A large number of software vendors provide powerful data mining tools. Some of these vendors are providers of only data mining and statistical analysis software, while others are large firms that provide wide ranges of software and hardware, along with consulting, in addition to data mining software products. Examples of the vendors that provide data mining tools include IBM (IBM SPSS Modeler, formerly known as SPSS PASW Modeler and Clementine), SAS (Enterprise Miner), StatSoft (Statistica Data Miner—now a Dell company), KXEN (Infinite Insight—now a SAP company), Salford (CART, MARS, TreeNet, and RandomForest), Angoss (KnowledgeSTUDIO and KnowledgeSeeker), and Megaputer (PolyAnalyst). Noticeably but not surprisingly, the most popular data mining tools were originally developed by the well-established statistical software companies (SPSS, SAS, and StatSoft). This is largely because statistics is the foundation of data mining, and these companies have the means to cost-effectively develop them into full-scale data mining systems.

Most of the business intelligence (BI) tool vendors (e.g., IBM Cognos, Oracle Hyperion, SAP Business Objects, Microstrategy, Teradata, Microsoft) also have some level of data mining capabilities integrated into their software offerings. These BI tools are still primarily focused on descriptive analytics in the sense of multidimensional modeling and data visualization and are not considered to be direct competitors of the data mining tool vendors.

In addition to the commercial data mining tools, there are several open source and/or free data mining software tools available over the Internet. Historically, the most popular free (and open source) data mining tool is Weka, which is developed by a number of researchers from the University of Waikato in New Zealand (which can be downloaded from cs.waikato.ac.nz/ml/weka/). Weka includes a large number of algorithms for different data mining tasks and has an intuitive user interface. Another recently released and quickly popularized free (for noncommercial use) data mining tool is RapidMiner, developed by RapidMiner.com (which can be downloaded from rapidminer.com). Its graphically enhanced user interface, use of a rather large number of algorithms, and incorporation of a variety of data visualization features set it apart from the rest of the other free data mining tools. Another free and open source data mining tool with an appealing graphical user interface is KNIME (which can be downloaded from knime.org).

The main difference between commercial tools, such as Enterprise Miner, IBM SPSS Modeler, and Statistica, and free tools, such as Weka, RapidMiner, and KNIME, is often the computational efficiency. The same data mining task involving a large data set may take a lot longer to complete with the free software, and for some algorithms, it may not even complete (i.e., it may crash due to the inefficient use of computer memory). Table 2.1 lists the major data mining products and their Websites.

Table 2.1 Popular Data Mining Software Tools

Product	Website (URL)
SAS Enterprise Miner	sas.com/technologies/bi/analytics/index.html
IBM SPSS Modeler	spss.com/Clementine
Statistica	statsoft.com/products/dataminer.htm
Intelligent Miner	ibm.com/software/data/iminer
PolyAnalyst	megaputer.com/polyanalyst.php
CART, MARS, TreeNet, and RandomForest	salford-systems.com
Insightful Miner	insightful.com
XLMiner	xlminer.net
KXEN Infinite Insight	kxen.com
GhostMiner	fqs.pl/ghostminer
Microsoft SQL Server Data Mining	microsoft.com/sqlserver/2012/data-mining.aspx
Knowledge Miner	knowledgeminer.net
Teradata Warehouse Miner	ncr.com/products/software/teradata_mining.htm
Oracle Data Mining (ODM)	otn.oracle.com/products/bi/9idmining.html

A suite of business intelligence capabilities that has become increasingly more popular for data mining studies is Microsoft's SQL Server. With this tool, data and models are stored in the same relational database environment, making model management a considerably easier task. The Microsoft Enterprise Consortium serves as the worldwide source for access to Microsoft's SQL Server 2012 software suite for academic purposes (i.e., teaching and research). The consortium has been established to enable universities around the world to access enterprise technology without having to maintain the necessary hardware and software on their own campuses. The consortium provides a wide range of business intelligence development tools (e.g., data mining, cube building, business reporting) as well as a number of large, realistic data sets from Sam's Club, Dillard's, and Tyson Foods. The Microsoft Enterprise Consortium is free of charge and can be used only for academic purposes. The Sam M Walton College of Business at the University of Arkansas hosts the enterprise system and allows consortium members and their students to access these resources using a simple remote desktop connection. The details about becoming a part of the consortium along with easy-to-follow tutorials and examples can be found at http://enterprise.waltoncollege.uark.edu/mec.asp.

In May 2014, kdnuggets.com (a well-known Web portal for data mining and analytics) conducted its 15th annual software poll on the following question: "What analytics, data mining, data science software/tools have you used in the past 12 months for a real project?" This poll received huge attention from the analytics and data mining community and vendors, attracting 3,285 voters. The poll measures both how widely a data mining tool is used and how strongly the vendors advocate for their tool. Here are some of the interesting findings that came out of the poll:

- Many data miners use more than one tool to carry out analytics projects. According to the poll, the average number of tools used by a person or vendor was 3.7 in 2014 (compared to 3.0 in 2013).
- The separation between commercial and free software continues to shrink. In 2014, 71% of voters used commercial software and 78% used free software. About 22% used only commercial software (down from 29% in 2013). About 28.5% used free software only (down only slightly from 30% in 2013). 49% used both free and commercial software (up from 41% in 2013). These numbers suggest that more users are including free, open source tools to their analytics arsenals.
- About 17.5% of voters report having used Hadoop or other Big Data tools (a measurable increase from 14% in 2013). This is an indicator of the increased popularity of Big Data tools and technologies.

The poll found these to be the top 10 tools, by share of users:

1. RapidMiner, 44.2% share (compared to 39.2% in 2013)
2. R, 38.5% (compared to 37.4% in 2013)
3. Excel, 25.8% (compared to 28.0% in 2013)
4. SQL, 25.3% (not available in 2013)
5. Python, 19.5% (compared to 13.3% in 2013)
6. Weka, 17.0% (compared to 14.3% in 2013)
7. KNIME, 15.0% (compared to 5.9% in 2013)
8. Hadoop, 12.7% (compared to 9.3% in 2013)
9. SAS base, 10.9% (compared to 10.7% in 2013)
10. Microsoft SQL Server, 10.5% (compared to 7.0% in 2013)

Among the tools with at least 2% share, the highest increase in 2014 was for Alteryx, which saw an increase of 1079%, to 3.1% in 2014 from 0.3% in 2013; SAP (including BusinessObjects, Sybase, and Hana), with an increase of 377%, to 6.8% from 1.4%; BayesiaLab, with an increase of 310%, to 4.1% from 1.0%; and KNIME, with an increase of 156%, to 15.0% from 5.9%. Among the tools with at least 2% share, the largest decline in 2014 was for StatSoft Statistica (now part of Dell), down 81%, to 1.7% share in 2014, from 9.0% in 2013 (partly due to lack of campaigning for Statistica, now that it is part of Dell); Stata, down 32%, to 1.4% from 2.1%; and BM Cognos, down 24%, to 1.8% from 2.4%.

Figure 2.4 shows the results of the poll for tools that received 100 or more votes. In the chart, the name of a tool is followed by the number of votes in parentheses and the "alone vote" percentage. The alone vote percentage is the percentage of voters who used only that tool, all by itself. For example, just 0.9% of Python users used only Python, while 35.1% of RapidMiner users indicated they used that tool alone.

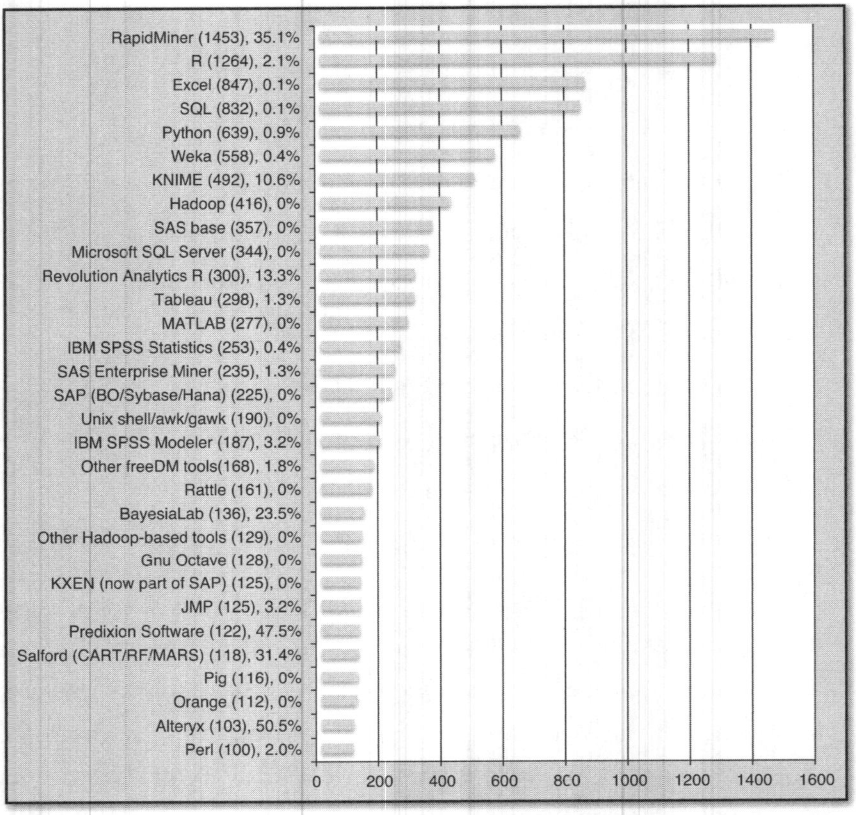

Figure 2.4 Most Popular Data Mining and Analytics Software Tools (User Poll Results)

Source: Used with permission of kdnuggets.com.

To reduce bias through multiple voting, in this poll, KDNuggets used email verification, which may potentially reduce the total number of votes but made the results less biased and more representative.

The Dark Side of Data Mining: Privacy Concerns

Data that is collected, stored, and analyzed in data mining often contains information about real people. Such information may include identification data (e.g., name, address, Social Security number, driver's license number, employee number), demographic data (e.g., age, sex, ethnicity, marital status, number of children), financial data (e.g., salary, gross family income, checking or savings account balance, home ownership, mortgage or loan account specifics, credit card limits and balances, investment account specifics), purchase history (i.e., what is bought from where and when, either from vendors' transaction records or from credit card transaction specifics), and other personal data (e.g., anniversary, pregnancy, illness, loss in the family, bankruptcy filings). Most of these data can be accessed through some third-party data providers. The main question here is the privacy of the person to whom the data belongs. In order to maintain the privacy and protection of individuals' right, data mining professionals have ethical and often legal obligations. One way to ethically handle private data is to de-identify customer records prior to applying data mining applications so that the records cannot be traced to an individual. Many publicly available data sources (e.g., CDC data, SEER data, UNOS data) are already de-identified. Prior to accessing these data sources, those mining data are often asked to consent that under no circumstances will they try to identify the individuals behind those figures.

In a number of instances in the recent past, companies have shared their customer data with others without seeking the explicit consent of their customers. For instance, as you might recall, in 2003, JetBlue Airlines provided more than a million passenger records of its customers to Torch Concepts, a U.S. government contractor. Torch then subsequently augmented the passenger data with additional information, such as family sizes and Social Security numbers—information purchased from a data broker called Acxiom. The consolidated personal database was intended to be used for a data mining project in order to develop potential terrorist profiles. All this was done without notification or consent of passengers. When news of the activities got out, however, dozens of privacy lawsuits were filed against JetBlue, Torch, and Acxiom, and several U.S. senators called for an investigation into the incident (Wald, 2004). Similar, but not as dramatic, privacy-related news came out in the recent past about popular social networking companies, which were allegedly selling customer-specific data to other companies for personalized targeted marketing.

Another peculiar story about data mining and privacy concerns made it to the headlines in 2012. In this instance, the company did not even use any private and/or personal data. Legally speaking, there was no violation of any laws. This story, which is about Target, goes as follows. In early 2012, an infamous story broke out about Target's practice of using predictive analytics. The story was about a teenager girl who was being sent advertising flyers and coupons by Target for the kind of things that a new mother-to-be would buy from a store like Target. The story goes like this: An angry man went into a Target outside Minneapolis, demanding to talk to a manager: "My daughter got this in the mail!" he said. "She's still in high school, and you're sending her coupons for baby clothes and cribs? Are you trying to encourage her to get pregnant?" The manager didn't have any idea what the man was talking about. He looked at the mailer. Sure enough, it was addressed to the man's daughter and contained advertisements for maternity clothing, nursery furniture, and pictures of smiling infants. The manager apologized and then called a few days later to apologize again. On the phone, though, the father was somewhat abashed. "I had a talk with my daughter," he said. "It turns out there's been some activities in my house I haven't been completely aware of. She's due in August. I owe you an apology."

As it turns out, Target figured out a teen girl was pregnant before her father did! Here is how they did it. Target assigns every customer a Guest ID number (tied to the person's credit card, name, or email address) that becomes a placeholder to keep a history of everything the person has bought. Target augments this data with any demographic information that it has collected from the customer or bought from other information sources. By using this type of information, Target had looked at historical buying data for all the women who had signed up for Target baby registries in the past. They analyzed the data from all directions, and soon, some useful patterns emerged. For example, lotions and special vitamins were among the products with interesting purchase patterns. Lots of people buy lotion, but Target noticed that women on the baby registry were buying larger quantities of unscented lotion around the beginning of their second trimester. Another analyst noted that sometime in the first 20 weeks, pregnant women loaded up on supplements like calcium, magnesium, and zinc. Many shoppers purchase soap and cotton balls, but when someone suddenly starts buying lots of scent-free soap and extra-big bags of cotton balls, in addition to hand sanitizers and washcloths, it signals that she could be getting close to her delivery date. Target was able to identify about 25 products that, when analyzed together, allowed them to assign each shopper a "pregnancy prediction" score. More importantly, Target could also estimate her due date to within a small window, so the company could send coupons timed to very specific stages of her pregnancy.

If you looked at this practice from a legality perspective, you would conclude that Target did not use any information that violates customers' privacy; rather, it used transactional data that almost every other retail chain is collecting and storing (and perhaps analyzing) about their customers. What was disturbing in this scenario was perhaps that pregnancy was the targeted concept. People tend to believe that certain events or concepts should be off-limits or treated extremely cautiously, such as terminal disease, divorce, and bankruptcy.

Application Example: Data Mining for Hollywood Managers

Predicting box office receipts (i.e., the financial success) of a particular motion picture is an interesting and challenging problem. According to some domain experts, the movie industry is the "land of hunches and wild guesses" due to the difficulty associated with forecasting product demand, and this makes the movie business in Hollywood a risky endeavor. In support of such observations, Jack Valenti (the longtime president and CEO of the Motion Picture Association of America) once mentioned that "no one can tell you how a movie is going to do in the marketplace...not until the film opens in a darkened theatre and sparks fly up between the screen and the audience." Entertainment industry trade journals and magazines have been full of examples, statements, and experiences that support this claim.

Like many other researchers who have attempted to shed light on this challenging real-world problem, Ramesh Sharda and Dursun Delen have been exploring the use of data mining to predict the financial performance of a motion picture at the box office before it even enters production (while the movie is nothing more than a conceptual idea). In their highly publicized prediction models, they convert the forecasting (or regression) problem into a classification problem; that is, rather than forecast the point estimate of box office receipts, they classify a movie based on its box office receipts in one of nine categories, ranging from "flop" to "blockbuster," making the problem a multinomial classification problem. Table 2.2 illustrates the definitions of the nine classes in terms of the range of box office receipts.

Data

Data was collected from a variety of movie-related databases (e.g., ShowBiz, IMDb, IMSDb, AllMovie) and consolidated into a single data set. The data set for the most recently developed models contained 2,632 movies released between 1998 and 2006. A summary of the independent variables along with their specifications is provided in Table 2.3. For more descriptive details and justification for inclusion of these independent variables, see Sharda and Delen (2006).

Table 2.2 Movie Classification, Based on Receipts

Class No.	1	2	3	4	5	6	7	8	9
Range (in millions)	< 1 (Flop)	> 1 < 10	> 10 < 20	> 20 < 40	> 40 < 65	> 65 < 100	> 100 < 150	> 150 < 200	> 200 (Blockbuster)

Table 2.3 Summary of the Predictive Variables

Independent Variable Name	Definition	Range of Possible Values
MPAA rating	The rating assigned by the Motion Picture Association of America (MPAA).	G, PG, PG-13, R, NR
Competition	Indicates the level at which each movie competes for the same pool of entertainment dollars against movies released at the same time.	High, Medium, Low
Star value	Signifies the presence of any box office superstars in the cast. A superstar actor/actress can be defined as one who contributes significantly to the up-front sale of the movie.	High, Medium, Low
Genre	Specifies the content category the movie belongs to. Unlike with the other categorical variables, a movie can be classified in more than one content category at the same time (e.g., action as well as comedy). Therefore, each content category is represented with a separate binary variable.	Sci-Fi, Epic Drama, Modern Drama, Thriller, Horror, Comedy, Cartoon, Action, Documentary
Special effects	Signifies the level of technical content and special effects (animations, sound, visual effects) used in the movie.	High, Medium, Low
Sequel	Specifies whether a movie is a sequel (value of 1) or not (value of 0).	Yes, No

Independent Variable Name	Definition	Range of Possible Values
Number of screens	Indicates the number of screens on which the movie is planned to be shown during its initial launch.	A positive integer

Methodology

Using a variety of data mining methods, including neural networks, decision trees, support vector machines, and three types of ensembles, Sharda and Delen developed the prediction models. The data from 1998 to 2005 were used as training data to build the prediction models, and the data from 2006 was used as the test data to assess and compare the models' prediction accuracy. Figure 2.5 shows the conceptual architecture for the proposed movie forecasting system, and Figure 2.6 shows the workflow of the model-building process.

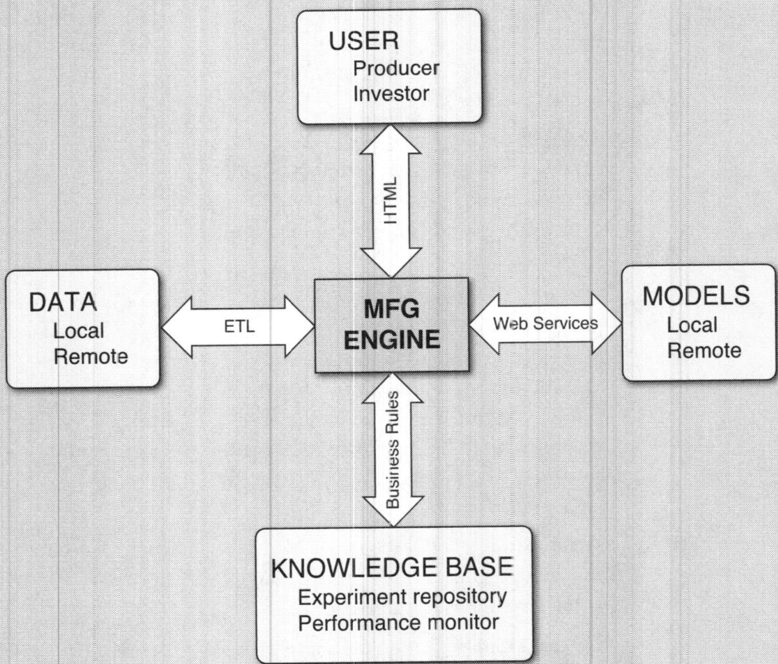

Figure 2.5 Conceptual Architecture for the Movie Prediction System

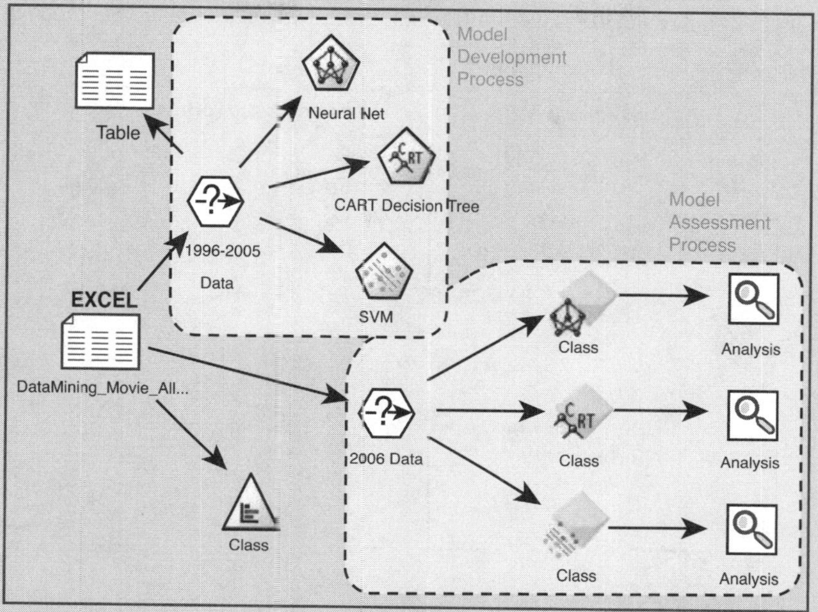

Figure 2.6 Workflow of the Model-Building Process

Results

Table 2.4 shows the prediction results of all three data mining methods, as well as the results of the three different ensembles. The first performance measure is the percentage correct classification rate, which is called bingo. The table also reports the 1-away correct classification rate (i.e., within one category). The results indicate that SVM performed the best among the individual prediction models, followed by ANN; the worst of the three was the CART decision tree algorithm. In general, the ensemble models performed better than the individual predictions models, and of those, the fusion algorithm performed the best. What is probably more important to decision makers, and stands out in the results table, is the significantly low standard deviation obtained from the ensembles compared to the individual models.

Table 2.4 Tabulated Prediction Results for Individual and Ensemble Models

	Prediction Models					
	Individual Models			**Ensemble Models**		
Performance Measure	SVM	ANN	C&RT	Random Forest	Boosted Trees	Information Fusion
Count (bingo)	192	182	140	189	187	194
Count (1-away)	104	120	126	121	104	120
Accuracy (% bingo)	55.49%	52.60%	40.46%	54.62%	54.05%	56.07%
Accuracy (% 1-away)	85.55%	87.28%	76.88%	89.60%	84.10%	90.75%
Standard deviation	0.93	0.87	1.05	0.76	0.84	0.63

Conclusions

The researchers claim that these prediction results are better than any others reported in the published literature for this problem domain. Beyond the attractive accuracy of their prediction results of the box office receipts, these models could also be used to further analyze (and potentially optimize) the decision variables in order to maximize the financial return. Specifically, the parameters used for modeling could be altered using the already trained prediction models in order to better understand the impact of different parameters on the end results. During this process, which is commonly referred to as *sensitivity analysis*, the decision maker of a given entertainment firm could find out, with a fairly high accuracy level, how much value a specific actor (or a specific release date, or the addition of more technical effects, etc.) brings to the financial success of a film, making the underlying system an invaluable decision aid. Figure 2.7 shows a screen shot from the Movie Forecast Guru software environment.

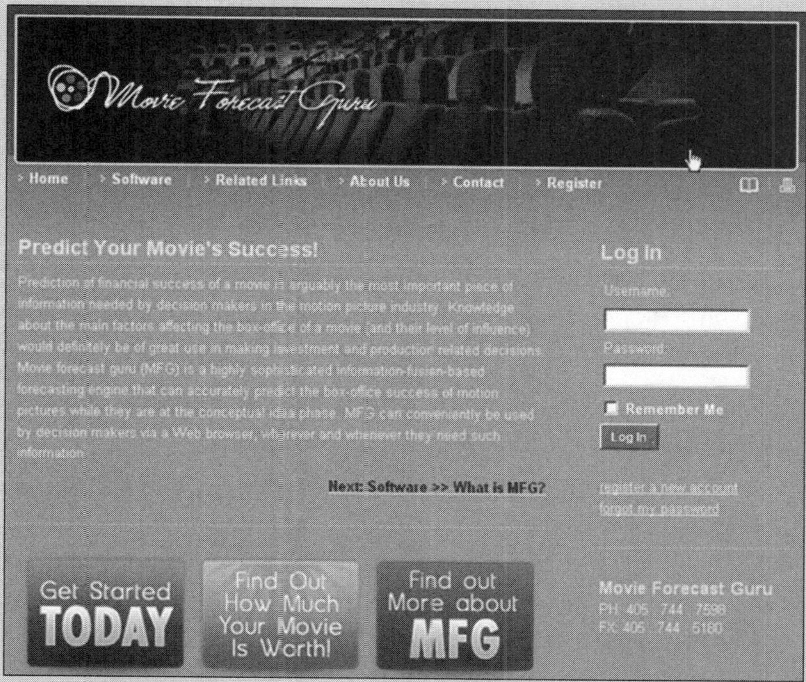

Figure 2.7 A Screen Shot of the Movie Forecast Guru

Sources: R. Sharda and D. Delen, "Predicting Box-Office Success of Motion Pictures with Neural Networks," *Expert Systems with Applications*, 30:243–254, 2006; D. Delen, R. Sharda, & P. Kumar, "Movie Forecast Guru: A Web-Based DSS for Hollywood Managers," *Decision Support Systems*, 43(4): 1151–1170, 2007.

References

CRISP-DM. (2013). *Cross-Industry Standard Process for Data Mining (CRISP-DM)*. www.the-modeling-agency.com/crisp-dm.pdf (accessed February 2013).

Davenport, T. H. (2006, January). "Competing on Analytics." *Harvard Business Review*.

Delen, D. (2009). "Analysis of Cancer Data: A Data Mining Approach," *Expert Systems*, 26(1): 100–112.

Delen, D., D. Cogdell, & N. Kasap. (2012). "A Comparative Analysis of Data Mining Methods in Predicting NCAA Bowl Outcomes," *International Journal of Forecasting*, 28: 543–552.

Delen, D., R. Sharda, & P. Kumar. (2007). "Movie Forecast Guru: A Web-Based DSS for Hollywood Managers." *Decision Support Systems*, 43(4): 1151–1170.

Delen, D., G. Walker, & A. Kadam. (2005). "Predicting Breast Cancer Survivability: A Comparison of Three Data Mining Methods," *Artificial Intelligence in Medicine*, 34(2): 113–127.

Fayyad, U., G. Piatetsky-Shapiro, & P. Smyth. (1996). "From Knowledge Discovery in Databases." *AI Magazine*, 17(3): 37–54.

Hill, K. (2012, February 16). "How Target Figured Out a Teen Girl Was Pregnant Before Her Father Did," *Forbes*.

Kohonen, T. (1982). "Self-Organized Formation of Topologically Correct Feature Maps." *Biological Cybernetics*, 43(1): 59–69.

Nemati, H. R., & C. D. Barko. (2001). "Issues in Organizational Data Mining: A Survey of Current Practices," *Journal of Data Warehousing*, 6(1): 25–36.

Nolan, R. (2012, February 21). "Behind the Cover Story: How Much Does Target Know?" *The New York Times*.

SEMMA. (2009). *SAS's Data Mining Process: Sample, Explore, Modify, Model, Assess.* http://sas.com/offices/europe/uk/technologies/analytics/datamining/miner/semma.html (accessed August 2009).

Sharda, R., & D. Delen. (2006). "Predicting Box-Office Success of Motion Pictures with Neural Networks," *Expert Systems with Applications*, 30: 243–254.

Wald, M. (2004). "U.S. Calls Release of JetBlue Data Improper." *The New York Times*, www.nytimes.com/2004/02/21/business/us-calls-release-of-jetblue-data-improper.html (accessed May 2014).

Wright, C. (2012). "Statistical Predictors of March Madness: An Examination of the NCAA Men's Basketball Championship." http://economics-files.pomona.edu/GarySmith/Econ190/Wright March Madness Final Paper.pdf (accessed October 2014).

3

The Data Mining Process

As has been the case in many other computational paradigms, extracting knowledge from large data repositories (i.e., data mining) started as trial-and-error experimental projects. Many practitioners have looked at the problem from the perspective of trying to characterize what works and what doesn't. For quite some time, data mining projects were carried out as rather artistic experimental endeavors. However, in order to methodically conduct data mining analysis, a standardized process needed to be developed and followed. Based on best practices, data mining researchers and practitioners have proposed several processes—workflows in the form of simple step-by-step methodologies—to maximize the likelihood of success in conducting data mining projects. These efforts have led to several standardized processes, some of which are described in this chapter.

The Knowledge Discovery in Databases (KDD) Process

One of the earliest and perhaps the first data mining process was the name of *knowledge discovery in databases* (*KDD*) methodology, proposed by Fayyad et al. (1996). In the KDD methodology, data mining is a single step where the patterns are extracted from data. They proposed KDD as a comprehensive end-to-end process that encompasses many individual steps to convert data into knowledge.

A pictorial representation of the KDD process is given in Figure 3.1. In the figure, the processing steps are shown as labeled directional arrows, and the result of each step is shown as a graphical image representing the artifact. As shown in Figure 3.1, the input to the KDD process is a collection of data coming from organizational databases and/or other external data sources. These data sources are often combined in a centralized data repository called a data warehouse. A data warehouse enables the KDD process to be implemented effectively and efficiently because it provides a single source for data to be mined. Once the data is consolidated in a unified data warehouse, the problem-specific data is extracted and prepared for further processing. Because the data is usually in a raw, incomplete, and dirty state, a through preprocessing needs to be conducted before the modeling can take place. Once the data is preprocessed and transformed into a form that's ready for modeling, a variety of modeling techniques are applied to the data to convert it into patterns, correlations, and predictive models. The discovered patterns must be validated, and then they need to be interpreted and internalized so that they can be converted into actionable information (i.e., knowledge). One important part of this process is the feedback loop that allows the process flow to redirect backward, from any step to any other previous steps, for rework and readjustments.

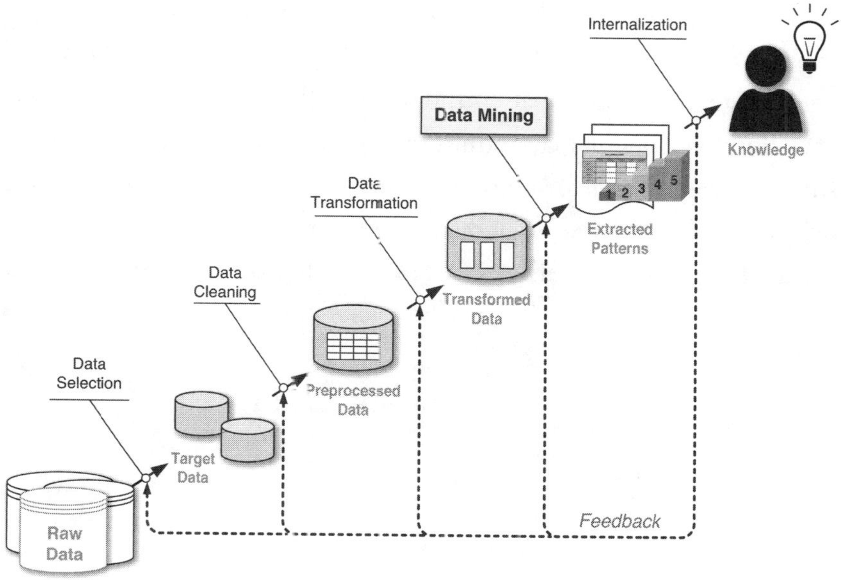

Figure 3.1 The Knowledge Discovery in Databases (KDD) Process

Cross-Industry Standard Process for Data Mining (CRISP-DM)

Another standardized data mining processes—arguably the most popular one—is called Cross-Industry Standard Process for Data Mining (CRISP-DM), which was proposed in the mid- to late 1990s by a European consortium of companies. It was intended to serve as a nonproprietary standard methodology for data mining projects (CRISP-DM, 2014). Figure 3.2 illustrates this six-step standardized process, which starts with a good understanding of the business problem and the need for the data mining project (i.e., the application domain) and ends with the deployment of the solution that satisfies the specific business need that started the data mining project in the first place. Even though these steps are shown in sequential order in the graphical representation, there usually is a great deal of backtracking. Because the data mining is driven by experience and experimentation, depending on the problem situation and the skills, knowledge, and experience of the analyst, the whole process can be very iterative (i.e., it may require going back and forth through the steps quite a few times) and time-intensive. Because each step is built on the outcome of the immediate predecessor, it's important to pay extra attention to the earlier steps in order to avoid putting an entire study on an incorrect path from the outset.

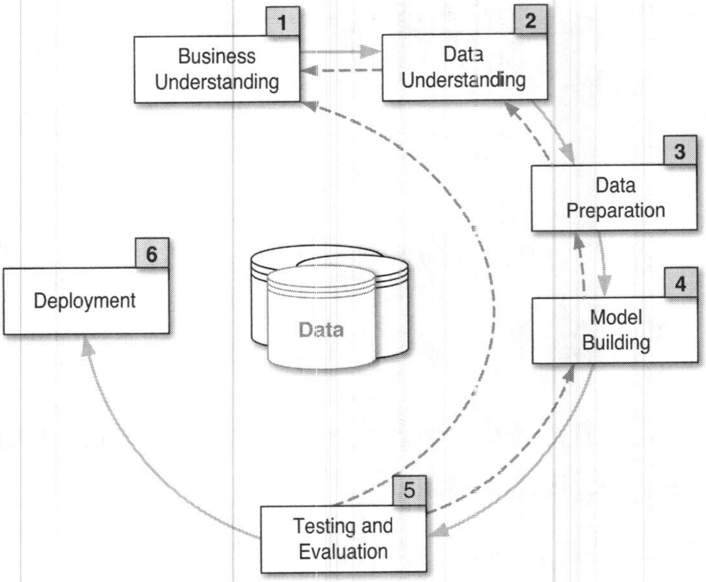

Figure 3.2 The CRISP-DM Data Mining Process

The following sections describe the steps of the CRISP-DM process in detail.

Step 1: Business Understanding

The key to success in any data mining project is to know what the study is for. Making this determination begins with gaining a thorough understanding of the managerial need for new knowledge and an explicit specification of the business objective regarding the study to be conducted. Specific goals such as "What are the common characteristics of the customers we have lost to our competitors recently?" or "What are typical profiles of our customers, and how much value does each of them provide to us?" are needed. Then a project plan for finding such knowledge is developed that specifies the people responsible for collecting the data, analyzing the data, and reporting the findings. At this early stage, a budget to support the study should also be established, at least at a high level and with rough numbers. For example, with a customer segmentation model developed for a retail catalog business, the identification of a business purpose meant identifying the type of customer that would be expected to yield a profitable return. A similar analysis would also be useful for credit card distributors. For business purposes, grocery stores often try to identify which items tend to be purchased together, so they can do better affinity positioning within the store or more intelligently guide promotional campaigns. Data mining has many useful business applications and can be applied to many business problems and opportunities. Intimately knowing the business purpose is critical to achieving success.

Step 2: Data Understanding

The second step in the CRISP-DM process is to make a perfect match between the business problem and the data being used to address it. That is, a data mining study is specific to a well-defined business task, and different business tasks require different sets of data. Therefore, it is important to identify the relevant data from many available data sources. Several key points must be considered in the data identification and selection process. First and foremost, an analyst should be clear and concise in describing the data mining task so that the most relevant data can be identified. For example, a retail data mining project might seek to identify spending behaviors of female shoppers who purchase seasonal clothes based on their demographics, credit card transactions, and socioeconomic attributes. Furthermore, the analyst should gain an intimate understanding of the data sources—for example, where the relevant data is stored and in what form, whether data collection is automated or happens manually, who collects the data, and how often the data are updated. The analyst should also understand variables by asking questions such as "What are the most relevant variables?" "Are there any synonymous and/or homonymous variables?" and "Are the variables independent of each other—that is, do they stand as a complete information source without overlapping or conflicting information?"

In order to better understand the data, the analyst often uses a variety of statistical and graphical techniques, such as simple statistical descriptors/summaries of each variable (e.g., for numeric variables, the average, minimum/maximum, median, and standard deviation are among the calculated measures, whereas for categorical variables the mode and frequency tables are calculated), correlation analysis, scatterplots, histograms, and box plots. Careful identification and selection of data sources and the most relevant variables can make it easier for data mining algorithms to quickly discover useful knowledge patterns.

Data sources for data selection can vary. Normally, data sources for business applications include demographic data (e.g., income, education, number of households, age), sociographic data (e.g., hobbies, club memberships, entertainment), and transactional data (e.g., sales record, credit card spending, issued checks), among others.

Data can be categorized as quantitative and qualitative. Quantitative data is measured using numeric values. It can be discrete (e.g., integers) or continuous (e.g., real numbers). Qualitative data, also known as categorical data, contains both nominal and ordinal data. Nominal data has finite non-ordered values. For example, gender data has two values: male and female. Ordinal data has finite ordered values. For example, customer credit ratings are considered ordinal data because the ratings can be excellent, fair, and bad. Quantitative data can be readily represented by some sort of probability distribution. A probability distribution describes how the data is dispersed and shaped. For instance, normally distributed data is symmetric and is commonly referred to as being a bell-shaped curve. Qualitative data may be coded to numbers and then described by frequency distributions. Once the relevant data is selected according to the data mining business objective, data preprocessing should be conducted. (For more details on data in data mining, see Chapter 4, "Data and Methods in Data Mining.")

Step 3: Data Preparation

The purpose of data preparation (commonly called data preprocessing) is to prepare the data identified in the previous step for analysis using data mining methods. Compared to the other steps in CRISP-DM, data preprocessing consumes the most time and effort—roughly 80% of the total time spent on a data mining project. This step requires such enormous effort because real-world data is generally incomplete (lacking attribute values, lacking certain attributes of interest, or containing only aggregate data), noisy (containing errors or outliers), and inconsistent (containing discrepancies in codes or names).

Some parts of the data may have different formats because they are taken from different data sources. The selected data may be from flat files, voice message, images, and Web pages, and it needs to be converted to a consistent and unified format. In general, data cleaning means filtering, aggregating, and filling in missing values (a.k.a. imputation). By filtering the data, an analyst examines the selected variables for outliers and redundancies. Outliers differ greatly from the majority of data, or data that is clearly out of range of the selected data groups. For example, if the age of a customer included in the data is 190, this must be a data entry error and should be identified and fixed (perhaps taken out of a data mining project that examines the various aspects of customers, since age is perceived to be a critical customer characteristic). Outliers may occur for many reasons, such as human errors or technical errors, or may naturally occur in a data set due to extreme events. Suppose the age of a credit card holder is recorded as "12." This is likely a data entry error, most likely made by a human. However, there might actually be an independently wealthy preteen with important purchasing habits. Arbitrarily deleting this outlier could dismiss valuable information.

Data may also be redundant, with the same information recorded in several different ways. Daily sales of a particular product are redundant to seasonal sales of the same product because an analyst can derive the sales from either daily data or seasonal data. Aggregating data reduces data dimensions. Note that although an aggregated data set has a small volume, the information remains. If a marketing promotion for furniture sales is considered in the next three or four years, the available daily sales data can be aggregated as annual sales data. The size of the sales data is then dramatically reduced. By smoothing data, missing values of the selected data are found and new or reasonable values are then added. These added values could be the average number of the variable (mean) or the mode. A missing value often causes no solution when a data mining algorithm is applied to discover the knowledge patterns.

Step 4: Model Building

In the fourth step of the CRISP-DM process, various modeling techniques are selected and applied to an already prepared data set in order to address the specific business need. The modeling step also encompasses the assessment and comparative analysis of the various types of models that can address the same type of data mining tasks (e.g., clustering, classification). Because there is not a universally known best method or algorithm for a specific data mining task, an analyst should use a variety of viable model types, along with a well-defined experimentation and assessment strategy, to identify the "best" method for a given data mining problem. Even for a single method or algorithm, a number of parameters need to be calibrated to obtain optimal results. Some methods may have specific requirements for the way the data is to be formatted; thus returning to the data preparation step is often necessary.

Depending on the business need, the data mining task can be a prediction (either classification or regression), an association, or a clustering/segmentation type. Each of these data mining tasks can use a variety of data mining methods and algorithms. For instance, classification-type data mining tasks can be accomplished by developing neural networks or by using decision trees, support vector machines, or logistic regression. These data mining methods and their respective algorithms are explained in Chapter 4 and Chapter 5, "Data Mining Algorithms."

The standard procedure for modeling in data mining is to divide a large preprocessed data set into subsets for training and validation or testing. Then the analyst can use a portion of the data (the training set) to develop the models (no matter what modeling technique and/or algorithm is used) and use the other portion of the data (the test set) for testing the model that was just built. The principle is that if you build a model on a particular set of data, it will of course test quite well on the data on which it was built. By dividing the data and using part of it for model development and testing it on a separate set of data, an analyst can create convincing and reliable results for the accuracy and reliability of the model. The idea of splitting the data into components is often carried to additional levels, with multiple splits in the practice of data mining. For further details about data splitting and other evaluation methods, see Chapter 4.

Step 5: Testing and Evaluation

In the fifth step of the CRISP-DM process, the developed models are assessed and evaluated for accuracy and generality. This step assesses the degree to which the selected model (or models) meets the business objectives and whether more models need to be developed and assessed. Another option is to test the developed model(s) in a real-world scenario if time and budget constraints permit. Even though the outcome of the developed models is expected to relate to the original business objectives, other findings that are not necessarily related to the original business objectives but that might also unveil additional information or hints for future directions are often discovered.

Testing and evaluation is a critical and challenging step. No value is added by the data mining task until the business value obtained from discovered knowledge patterns is identified and recognized. Determining the business value from discovered knowledge patterns is somewhat similar to playing with puzzles. The extracted knowledge patterns are pieces of the puzzle that need to be put together in the context of the specific business purpose. The success of this identification operation depends on the interaction among data analysts, business analysts, and decision makers (e.g., business managers). Data analysts may fully understand the data mining objectives and what they mean to the business, and business analysts and decision makers may not have the technical knowledge to interpret the results of sophisticated mathematical solutions; therefore, interaction among them is necessary. In order to properly interpret knowledge patterns, it is often necessary to use a variety of tabulation and visualization techniques (e.g., pivot tables, cross-tabulation of findings, pie charts, histograms, box plots, scatterplots).

Step 6: Deployment

Development and assessment of models is not the end of a data mining project. Even if the purpose of a model is to do a simple exploration of the data, the knowledge gained from such exploration needs to be organized and presented in a way that the end user can understand and benefit from. Depending on the requirements, the deployment phase can be as simple as generating a report or as complex as implementing a repeatable data mining process across the enterprise. In many cases, it is the customer, not the data analyst, who carries out the deployment steps. However, even if an analyst will not carry out the deployment effort, it is important for the customer to understand up front what actions need to be carried out in order to actually make use of the created models.

The deployment step of the CRISP-DM process may also include maintenance activities for the deployed models. Because the business is constantly changing, the data that reflects the business activities is also changing. Over time, the models (and the patterns embedded within them) built on the old data may become obsolete, irrelevant, or misleading. Therefore, monitoring and maintaining the models are important if the data mining results are to become a part of the day-to-day business and its environment. Careful preparation of a maintenance strategy helps avoid unnecessarily long periods of incorrect use of data mining results. In order to monitor the deployment of the data mining result(s), a project needs a detailed plan for the monitoring process, which may not be trivial with complex data mining models.

The CRISP-DM process is the most complete and most popular data mining methodology practice in industry as well as in academia. Rather than use it as is, practitioners add their own insights to make it specific to their style of practice.

SEMMA

In order to be applied successfully, a data mining solution must be viewed as a process rather than as a set of tools or techniques. In addition to KDD and CRISP-DM, there is yet another well-known methodology, called SEMMA, developed by the SAS Institute. The acronym SEMMA stands for **s**ample, **e**xplore, **mo**dify, **m**odel, **a**ssess. Beginning with a statistically representative sample of data, SEMMA intends to make it easy to apply exploratory statistical and visualization techniques, select and transform the most significant predictive variables, model the variables to predict outcomes, and confirm the model's accuracy. A pictorial representation of SEMMA is given in Figure 3.3.

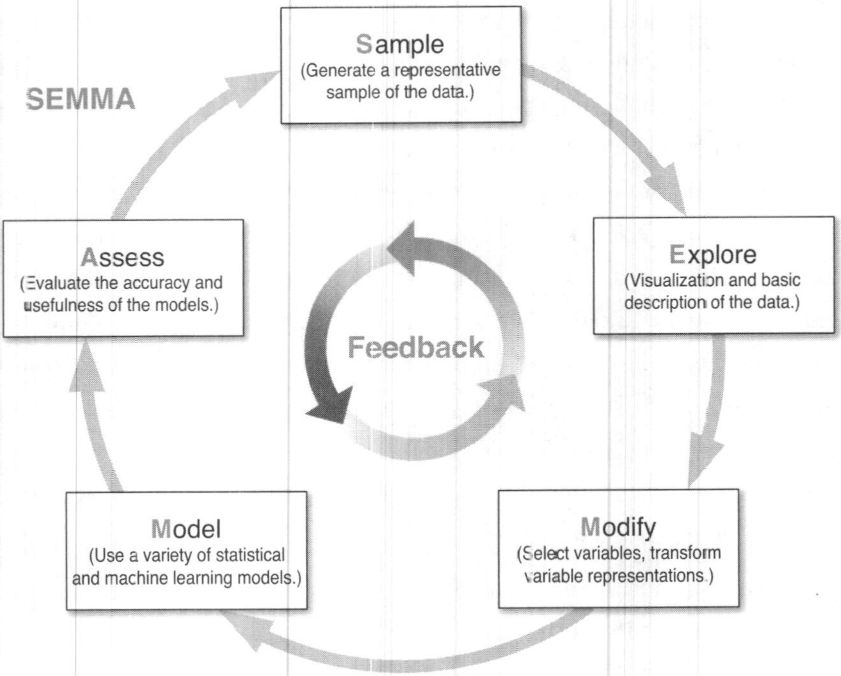

Figure 3.3 Schematic of SAS Institute's SEMMA

By assessing the outcome of each stage in the SEMMA process, an analyst can determine how to model new questions raised by the previous results and thus return to earlier steps for additional refinement of the data. Like CRISP-DM, SEMMA is also driven by a highly iterative experimentation cycle. The following sections describe the five steps of the SEMMA process.

Step 1: Sample

In the sample step, a portion of a large data set (big enough to contain significant information yet small enough to manipulate quickly) is extracted. For optimal cost and computational performance, some (including the SAS Institute) advocate using a sampling strategy that involves applying a reliable, statistically representative sample of the full-detail data. In the case of very large data sets, mining a representative sample instead of the whole volume may drastically reduce the processing time required to get crucial business information. If general patterns appear in the data as a whole, they will be traceable in a representative sample. If a niche (a rare pattern) is so tiny that it is not represented in a sample and yet so important that it influences the big picture, it should be discovered using exploratory data description methods. It is also a good idea to create partitioned data sets for better accuracy assessment.

The sample step of the SEMMA process involves three substeps:

- **Training.** Used for model fitting.
- **Validation.** Used for assessment and to prevent overfitting.
- **Test.** Used to obtain an honest assessment of how well a model generalizes.

For a more detailed discussion and relevant techniques for assessment and validation of data mining models, see Chapter 4.

Step 2: Explore

During the second step of the SEMMA process, the analyst searches for unanticipated trends and anomalies in order to gain a better understanding of the data set. The analyst explores the data visually or numerically for inherent trends or groupings. Exploration helps refine and redirect the discovery process. If visual exploration does not reveal clear trends, the analyst can explore the data by using statistical techniques, including factor analysis, correspondence analysis, and clustering. For example, in data mining for a direct mail campaign, clustering might reveal groups of customers with distinct ordering patterns. Limiting the discovery process to each of these distinct groups individually may increase the likelihood of exploring richer patterns that may not be strong enough to be detected if the whole data set is to be processed together.

Step 3: Modify

During the third step of the SEMMA process, the analyst creates, selects, and transforms the variables on which to focus the model construction process. Based on the discoveries in the exploration phase, the analyst may need to manipulate data to include information such as the grouping of customers and significant subgroups or to introduce new variables. It may also be necessary to look for outliers and reduce the number of variables in order to narrow them down to the most significant ones. The analyst may also need to modify data when data that is mined changes. Because data mining is a dynamic, iterative process, an analyst can update data mining methods or models when new information is available.

Step 4: Model

The fourth step of the SEMMA process is where the analyst searches for a variable combination that reliably predicts a desired outcome. The goal is to construct models that explain patterns in the data. Modeling techniques in data mining include artificial neural networks, decision trees, rough set analysis, support vector machines, logistic models, and other statistical models, such as time-series analysis, memory-based reasoning, and principal component analysis. Each type of model has particular strengths and is appropriate in specific data mining situations and for different types of data. For example, artificial neural networks are very good at fitting highly complex non-linear relationships, while rough set analysis is known to produce reliable results with uncertain and imprecise problem situations.

Step 5: Assess

In the final step of the SEMMA process, the analyst evaluates the usefulness and reliability of the findings from the data mining process. This is basically an estimation of how well the model performs. A common way to assess a model is to apply it to a portion of a data set that was put aside during the sampling stage and not used during model building. If the model is valid, it should work for this reserved sample as well as for the sample used to construct the model. Similarly, it is possible to test the model against known data. For example, if you know which customers in a file had high retention rates and your model predicts retention, you can check to see whether the model selects those customers accurately. In addition, practical applications of the model, such as partial mailings in a direct mail campaign, help prove the validity of a model.

SEMMA Versus CRISP-DM

The SEMMA approach is quite compatible with the CRISP-DM approach. Both aim to streamline the knowledge discovery process. Both were created as broad frameworks that should be adapted to specific circumstances. In both, once models are obtained and tested, they can then be deployed to gain value with respect to business or research application. Even though they have the same goal and are similar, SEMMA and CRISP-DM have a few differences. Table 3.1 presents these differences.

Table 3.1 Comparison of SEMMA and CRISP-DM

Task	CRISP-DM	SEMMA	Comments
Project initiation	Business understanding	—	In this phase, CRISP-DM includes activities like project initiation, problem definition, and goal setting. SEMMA does not have a step for this phase.
Data access	Data understanding	Sample Explore	In this phase, both CRISP-DM and SEMMA have steps to access, sample, and explore the data.
Data transformation	Data preparation	Modify	In this phase, both CRISP-DM and SEMMA process the data to make it amenable to machine processing.
Model building	Model building	Model	In this phase, both CRISP-DM and SEMMA build and test various models.
Project evaluation	Testing and evaluation	Assess	In this phase, both CRISP-DM and SEMMA assess the findings against the project goals.
Project finalization	Deployment	—	CRISP-DM prescribes deploying the results, while SEMMA does not explicitly have a step for this.

Six Sigma for Data Mining

Six Sigma is a popular business management philosophy that focuses on reducing deviations (i.e., sigma) from perfection by rigorously and systematically using proven quality control principles and techniques. This hugely popular management philosophy was first introduced by Motorola in the 1980s in the context of manufacturing management. Since then it has been adopted by many companies and organizations in a wide verity of business contexts beyond manufacturing. Ideally, Six Sigma promotes zero defects and zero tolerance, which can roughly be translated in the business context as error-free perfect business execution. Six Sigma methodology has manifested itself in the business world with DMAIC, which involves five steps: **D**efine, **M**easure, **A**nalyze, **I**mprove, and **C**ontrol.

Because of its success in many other business problems and settings, DMAIC methodology is also applied to data mining projects. Figure 3.4 shows the DMAIC methodology as a simplified flow diagram. The following sections describe these steps.

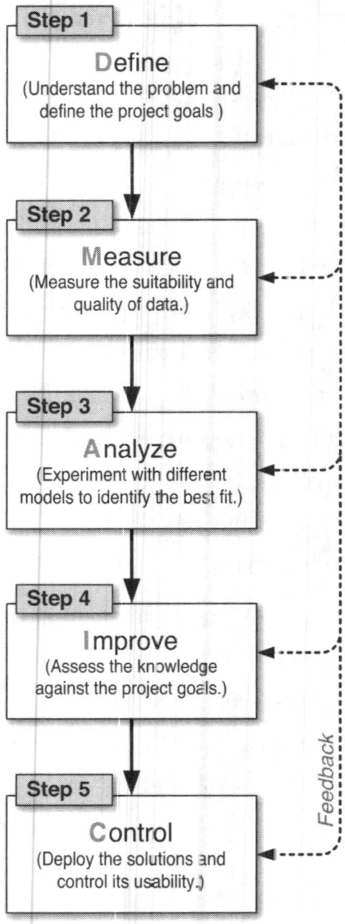

Figure 3.4 Six Sigma DMAIC Methodology

Step 1: Define

The first step in the DMAIC process is about getting the project set up and started. It involves several steps: (1) gaining a thorough understanding of the business needs, (2) identifying the most pressing problem, (3) defining goals and objectives, (4) identifying and defining the data and other resources needed to investigate the business problem, and (5) developing a detailed project plan. As you may have noticed, there is a significant overlap between this step and the first step in the CRISP-DM process.

Step 2: Measure

The second step of the DMAIC process involves assessing the mapping between organizational data repositories and the business problem. Since data mining requires relevant, clean, and usable data, identification and creation of such resources is of critical importance to the success of a data mining project. In this step, the identified data sources are consolidated and transformed into a format that is amenable to machine processing.

Step 3: Analyze

The third step of the DMAIC process is to use a series of data mining techniques to develop models. There are many possible techniques, and most of them are machine learning techniques with many parameters to optimize. There is not a single best technique, so an analyst needs to apply several probable techniques and experiment with them to identify and develop the most appropriate model.

Step 4: Improve

The fourth step of the DMAIC process is to investigate the improvement possibilities. Improvements can be at the technique level or at the business problem level. For instance, if the model results are not satisfactory, other more sophisticated techniques (e.g., ensemble systems) can be used to boost the performance of the models. Also, if the modeling results are not clearly addressing the business problem, the analyst can loop back to previous steps to reexamine and improve the structure of the analysis. The analyst can also further investigate and restate the business problem.

Step 5: Control

The last step of the DMAIC process is to assess the project outcomes, and if they are found to be satisfactory, the models and result are disseminated to decision makers and/or integrated into the existing business intelligence systems for automation.

Six Sigma–based DMAIC methodology resembles the CRISP-DM process. There is no evidence to suggest that one was inspired by the other. However, because these two methodologies involve rather logical and straightforward steps in any business system analysis effort, they need not have to be inspired one another. The recumbence might simply be a coincidence. Users of DMAIC are rarely compared to CRISP-DM because these two methodologies are used to address different types of business applications. Many organizations around the world are using DMAIC and CRIPS-DM and their variants to conduct process improvement and data mining projects.

Which Methodology Is Best?

Even though some data mining processes are more elaborate than the others, there is no sure way to compare them. They all have pros and cons. Some are more problem focused, while others are more analysis driven. A business that does data mining adopts one of these methodologies and then is likely to slightly modify it to make it fit the business or the data. KDNuggets, a well-known and well-respected Web portal for data mining, conducted a survey in which it asked "Which one is the best?" Figure 3.5 shows the polling results of that survey (KDNuggets, 2007).

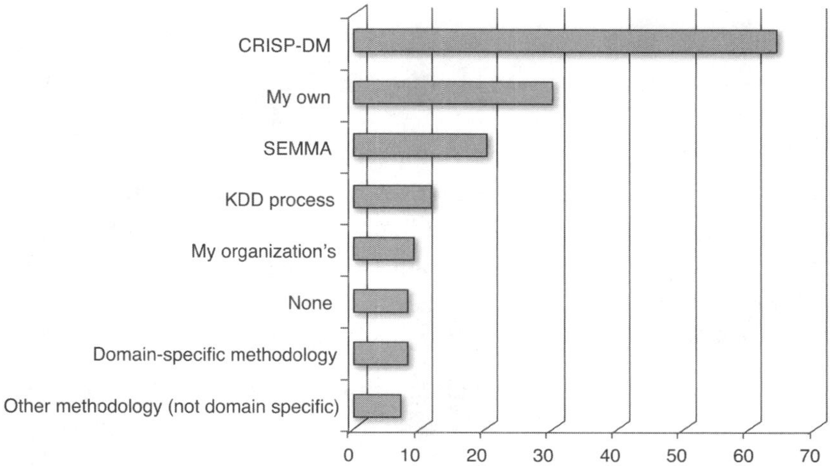

Figure 3.5 A Poll on Popular Data Mining Methodologies

Source: Used with permission from kdnuggets.com

As the survey results indicate, CRISP-DM dominates as the most popular data mining process. In addition, many of the ones that fall under "My own" are also customizations of CRISP-DM. Compared to the other methodologies, CRISP-DM is the most complete and most mature process for data mining.

An Exemplary Data Mining Case: Mining Cancer Data for New Knowledge

Cancer is one of the most deadly diseases. According to the American Cancer Society, half of all men and one-third of all women in the United States will develop cancer during their lifetimes, and approximately 1.5 million new cancer cases will be diagnosed in 2014. Cancer is the second-most-common cause of death in the United States and in the world, exceeded only by cardiovascular disease. This year, more than 500,000 Americans are expected to die of cancer—more than 1,300 people a day—accounting for nearly 1 of every 4 deaths.

Cancer is a group of diseases generally characterized by uncontrolled growth and spread of abnormal cells. If that growth and/or spread is not controlled, it can result in death. Cancer is believed to be caused by both external factors (e.g., tobacco use, infectious organisms, exposure to chemicals or radiation) and internal factors (e.g., inherited mutations, hormones, immune conditions, mutations that occur due to metabolism). These causal factors may act together or in sequence to initiate or promote carcinogenesis. Cancer is treated with surgery, radiation, chemotherapy, hormone therapy, biological therapy, and targeted therapy. Survival statistics vary greatly by cancer type and stage at diagnosis.

The five-year relative survival rate for all cancers is improving. Between 1991 and 2013, the cancer mortality rate declined by 20%, translating to the avoidance of about 1.2 million deaths during that timeframe. That's more than 400 lives saved per day! The improvement in survival reflects progress in diagnosing certain cancers at an earlier stage and improvements in treatment. Further improvements are needed to prevent and treat cancer.

Even though cancer research has traditionally been clinical and biological in nature, in recent years, data-driven analytic studies have become a common complement. In medical domains where data- and analytics-driven research have been applied successfully, novel research directions have been identified to further advance clinical and biological studies. Using various types of data, including molecular, clinical, literature-based, and clinical-trial data, along with suitable data mining tools and techniques, researchers have been able to identify novel patterns, paving the road toward a cancer-free society.

In one study, Delen (2009) used three popular data mining techniques—decision trees, artificial neural networks, and support vector machines—in conjunction with logistic regression to develop prediction models for prostate cancer survivability. The data set contained around 120,000 records and 77 variables. A k-fold cross-validation methodology was used in model building, evaluation, and comparison. The results showed that support vector machines, which are some of the most popular prediction algorithms, were the most accurate predictors (with a test set accuracy of 92.85%)

for this domain, followed by artificial neural networks and decision trees. Furthermore, using a sensitivity analysis–based evaluation method, the study also revealed novel patterns related to prognostic factors related to prostate cancer.

In a related study, Delen et al. (2005) used two data mining algorithms—artificial neural networks and decision trees—and a statistical classification algorithm—logistic regression—to develop prediction models for breast cancer survival using a large data set (more than 200,000 cases). Using a 10-fold cross-validation method to measure the unbiased estimate of the prediction models for performance comparison purposes, the results indicated that the decision tree (C5 algorithm) was the best predictor, with 93.6% accuracy on the holdout sample (which was the best prediction accuracy reported in the literature), followed by artificial neural networks, with 91.2% accuracy, and logistic regression, with 89.2% accuracy. A tabulation of the comparative precision results for all model types is shown in Figure 3.6. Further analysis of prediction models revealed prioritized importance of the prognostic factors, which can then be used as basis for further clinical and biological research studies.

These examples (among many others in the medical literature) show that advanced data mining techniques can be used to develop models that possess a high degree of predictive as well as explanatory power. Although data mining methods are capable of extracting patterns and relationships hidden deep in large and complex medical databases, without cooperation and feedback from medical experts, their results are not of much use. The patterns found via data mining methods should be evaluated by medical professionals who have years of experience in the problem domain to decide whether they are logical, actionable, and novel enough to warrant new research directions. In short, data mining is not meant to replace medical professionals and researchers but to complement their invaluable efforts to provide data-driven new research directions and to ultimately save more human lives.

Fold No	Neural Networks (MLP)					Decision Tree Induction (C5)					Logistic Regression				
	Confusion Matrix		Accuracy	Sensitivity	Specificity	Confusion Matrix		Accuracy	Sensitivity	Specificity	Confusion Matrix		Accuracy	Sensitivity	Specificity
1	7571	844	0.9165	0.9535	0.8719	7828	587	0.9363	0.9586	0.9078	7672	743	0.8912	0.9015	0.8766
	369	5747				338	5778				838	5277			
2	7589	729	0.9271	0.9578	0.8905	7737	581	0.9403	0.9639	0.9113	7543	773	0.8906	0.9018	0.8756
	334	5926				290	5970				821	5439			
3	7567	768	0.9214	0.9537	0.8818	7741	594	0.9356	0.9584	0.9065	7602	732	0.8915	0.9011	0.8779
	367	5730				336	5761				834	5261			
4	7508	796	0.9169	0.9480	0.8798	7703	601	0.9356	0.9582	0.9076	7607	696	0.8951	0.9017	0.8860
	412	5824				336	5900				829	5407			
5	7609	809	0.9239	0.9549	0.8903	7789	629	0.9348	0.9607	0.9021	7659	757	0.8908	0.9022	0.8747
	359	6565				319	5796				830	5284			
6	7390	908	0.8914	0.9179	0.8581	7694	604	0.9363	0.9604	0.9062	7554	743	0.8900	0.8992	0.8771
	661	5491				317	5835				847	5305			
7	7298	751	0.9081	0.9290	0.8823	7464	585	0.9374	0.9605	0.9095	7333	716	0.8949	0.9037	0.8831
	558	5631				307	5882				781	5408			
8	7069	977	0.9024	0.9442	0.8565	7436	610	0.9353	0.9594	0.9068	7269	773	0.8894	0.9001	0.8756
	418	5832				315	5935				807	5443			
9	7290	958	0.9040	0.9454	0.8559	7621	627	0.9360	0.9631	0.9027	7501	747	0.8923	0.9036	0.8767
	421	5691				292	5820				800	5310			
10	7475	764	0.9098	0.9330	0.8809	7625	614	0.9349	0.9591	0.9052	7518	716	0.8938	0.9022	0.8824
	537	5651				325	5863				815	5372			
Mean			0.9121	0.9437	0.8748			0.9362	0.9602	0.9066			0.8920	0.9017	0.8786
St. Dev.			0.0111	0.0131	0.0135			0.0016	0.0019	0.0028			0.0020	0.0014	0.0038

Confusion matrix shows the classification of the cases in the test dataset
In Confusion Matrix, the columns denote the actual cases and the rows denote the predicted
Accuracy = (TP + TN) / (TP + FP + TN + FN); Sensitivity = TP / (TP + FN); Specificity = TN / (TN + FP)

Figure 3.6 Tabular Results for a 10-Fold Cross-Validation for All Folds and All Model Types

References

CRISP-DM. (2014). "Cross-Industry Standard Process for Data Mining (CRISP-DM)." www.the-modeling-agency.com/crisp-dm.pdf (accessed May 22, 2014).

Delen, D. (2009). "Analysis of Cancer Data: A Data Mining Approach," *Expert Systems*, 26(1): 100–112.

Delen, D., & N. Patil. (2006, January). "Knowledge Extraction from Prostate Cancer Data," *Proceedings of the 39th Annual Hawaii International Conference on Systems Sciences*, 5: 92b.

Delen, D., G. Walker, & A. Kadam. (2005). "Predicting Breast Cancer Survivability: A Comparison of Three Data Mining Methods," *Artificial Intelligence in Medicine*, 34(2): 113–127.

Fayyad, U., G. Piatetsky-Shapiro, and P. Smyth. (1996). "From Knowledge Discovery in Databases." *AI Magazine*, 17(3): 37–54.

KDNuggets. (2007). "Poll: What Main Methodology Are You Using for Data Mining?" www.kdnuggets.com/polls/2007/data_mining_methodology.htm (accessed September 2014).

4

Data and Methods in Data Mining

The popularity of data mining can be attributed to its ability to discover hidden patterns deep inside large data repositories that organizations are automatically accumulating these days. Data by itself is not valuable to organizations that are desperately looking for ways to make evidence-based intelligent decisions. It is only after methods are applied to data that the invaluable knowledge organizations need can be brought out. A large variety of methods and algorithms can be used to carry out different types of data mining tasks. Each of these methods has advantages and disadvantages for a given project, depending on the specific characteristics of the problem being addressed and the data set being used. Identifying the best method to use usually requires quite a bit of experimentation and sound comparative assessment.

This chapter provides rather detailed yet nontechnical descriptions of these methods and commonly applied assessment techniques. Before we get to that, though we need to first talk about data as it relates to data mining.

The Nature of Data in Data Mining

Data refers to a collection of facts usually obtained as a result of experiences, observations, or experiments. Data may consist of numbers, letters, words, images, voice recordings, and other variables.

Data is often viewed as the lowest level of abstraction from which information and then knowledge are derived.

At the highest level of abstraction, one can classify data as structured and unstructured (or semi-structured). Unstructured/semi-structured data is composed of any combination of textual, imagery, voice, and Web content. Unstructured/semi-structured is covered in more detail in Chapter 6, "Text Analytics and Sentiment Analysis." Structured data is the data that data mining algorithms uses, and it can be classified as categorical or numeric. Categorical data can be subdivided into nominal or ordinal data, whereas numeric data can be subdivided into interval or ratio data. Figure 4.1 shows a simple taxonomy of data as it relates to data mining.

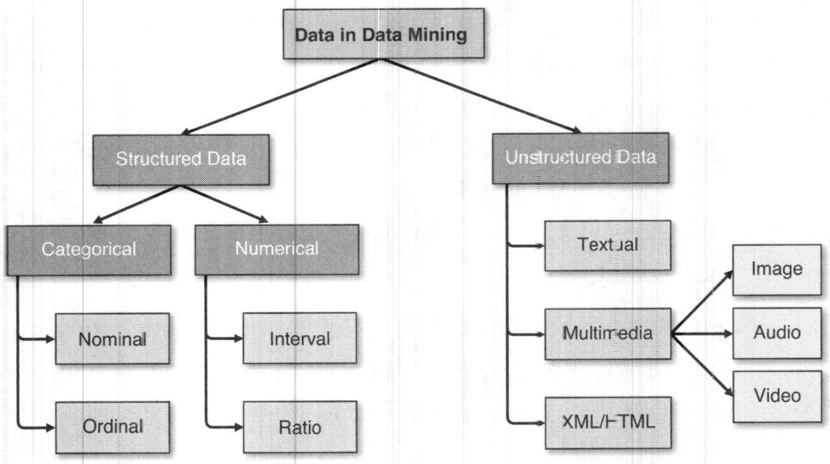

Figure 4.1 A Simple Taxonomy of Data in Data Mining

Categorical data falls into multiple classes used to divide a variable into specific groups. Examples of categorical variables include race, sex, age group, and educational level. Although age group and educational level may also be considered in a numerical manner by using exact values for age and highest grade completed, it is often more informative to categorize such variables into a relatively small number of ordered classes. The categorical data may also be called *discrete data*, implying that it represents a finite number of values with no continuum between them. Even if the values used for the categorical (or discrete) variables are numeric, these numbers are nothing more than symbols and do not imply the possibility of calculating fractional values.

Nominal data includes measurements of simple codes assigned to objects as labels, which are not measurements. For example, the variable marital status can be generally categorized as (1) single, (2) married, or (3) divorced. Nominal data can be represented with binomial values having two possible values (e.g., yes/no, true/false, good/bad), or multinomial values having three or more possible values (e.g., brown/green/blue, white/black/Latino/Asian, single/married/divorced).

Ordinal data includes codes assigned to objects or events as labels that also represent the rank order among them. For example, the variable credit score can be generally categorized as (1) low, (2) medium, or (3) high. Similar ordered relationships can be seen in variables such as age group (i.e., child/young/middle-aged/elderly) and educational level (i.e., high school/college/graduate school). Some data mining algorithms, such as ordinal multiple logistic regression, take into account this additional rank order information to build a better classification model.

Numeric data represents the numeric values of specific variables. Examples of numerically valued variables include age, number of children, total household income (in U.S. dollars), travel distance (in miles), and temperature (in Fahrenheit degrees). Numeric values representing a variable can be integer (taking only whole numbers) or real (taking also the fractional number). Numeric data may also be called *continuous data*, implying that the variable contains continuous measures on a specific scale that allows insertion of interim values. Unlike a discrete variable, which represents finite, countable data, a continuous variable represents scalable measurements, and it is possible for the data to contain an infinite number of fractional values.

Interval data includes variables that can be measured on interval scales. A common example of interval scale measurement is temperature on the Celsius scale. In this particular scale, the unit of measurement is 1/100 of the difference between the melting temperature and the boiling temperature of water in atmospheric pressure; that is, there is not an absolute zero value.

Ratio data includes measurement variables commonly found in the physical sciences and engineering. Mass, length, time, plane angle, energy, and electric charge are examples of physical measures that are ratio data. The scale type takes its name from the fact that measurement is the estimation of the ratio between a magnitude of a continuous quantity and a unit magnitude of the same kind. Informally, the distinguishing feature of a ratio scale is the possession of a nonarbitrary zero value. For example, the Kelvin temperature scale has a nonarbitrary zero point of absolute zero, which is equal to –273.15 degrees Celsius. This zero point is nonarbitrary because the particles that comprise matter at this temperature have zero kinetic energy.

Other data types, including textual, multimedia (e.g., image, audio, video), and XML/HTML need to be converted into some form of categorical or numeric representation before they can be processed by data mining algorithms. (For more details about how to mine knowledge from unstructured and semi-structured data, see Chapter 6.) Data can also be geographically oriented, so that the numeric and nominal data is enriched with locational/geographic information; this is often referred to as spatial data. Furthermore, regardless of its structure, data can also be classified as static or dynamic (i.e., temporal or time-series), based on its relationship to time.

Some data mining methods and algorithms are very particular about the type of data they can handle. Providing them with incompatible data types may lead to incorrect models or (more often) halt the model development process. For example, some data mining methods (e.g., neural networks, support vector machines, logistic regression) need all the variables (both input and output) represented as numerically valued variables. The nominal or ordinal variables are converted into numeric representations using some type of 1-of-*N* pseudo variables (e.g., a categorical variable with three unique values may be transformed into three pseudo variables with binary values—1 or 0). Because this process may increase the number of variables, one should be cautious about the effect of such representations, especially for the categorical variables that have large numbers of unique values.

Similarly, some data mining methods, such as ID3 (a classic decision tree algorithm) and rough sets (a relatively new rule-induction algorithm), need all the variables represented as categorically valued variables. Early versions of these methods required the user to discretize numeric variables into categorical representations before processing them with the algorithm. The good news is that most implementations of these algorithms in widely available software tools accept a mix of numeric and nominal variables and internally make the necessary conversions before processing the data.

Preprocessing of Data for Data Mining

The quality and usefulness of models created using data mining methods depends largely on the quality of the data used to build them. In that sense, perhaps the GIGO rule (i.e., garbage-in, garbage-out) is more applicable to data mining than to any other application domain. Therefore, the purpose of data preparation (or more commonly called *data preprocessing*) is to eliminate the possibility of GIGO errors by methodically cleaning and transforming data into a standard form so that it can be properly analyzed by appropriate data mining methods. Compared to the other activities in the data mining process, data preprocessing demands and consumes the most time and effort, often accounting for more than 80% of the total time spent on a project. Preprocessing requires such an enormous effort because real-world data is generally incomplete (lacking attribute values, lacking certain attributes of interest, or containing only aggregate data), dirty/noisy (containing errors or outliers), and inconsistent (containing discrepancies in codes or names). Therefore, diligent effort is necessary to clean and transform the data into a form that can be mined while making sure that the preprocessing does not bias the data; it is important to preserve the inherent patterns and relationships of the data. Figure 4.2 shows the four-step systematic process commonly followed to convert the raw real-world data into well-formed data sets that can be mined.

In the first phase of data preprocessing, the relevant data is collected from the identified sources, the necessary records and variables are selected (filtering out the unnecessary data, based on an intimate understanding of the data), and the records coming from multiple data sources are integrated. During this phase, domain-specific understanding of the data is applied to properly handle the synonyms and homonyms in the data.

In the second phase of data preprocessing, known as data scrubbing, the data is cleaned. In this step, the values in the data set are identified and dealt with. In some cases, missing values are an anomaly in the data set, in which case they need to be imputed (i.e., filled with a most probable value) or ignored; in other cases, the missing values are a natural part of the data set (e.g., the household income field is often left unanswered by people who are in the top income tier). In this step, the analyst should also identify noisy values in the data (i.e., outliers) and smooth them out. In addition, inconsistencies (unusual values within a variable) in the data should be handled using domain knowledge and/or expert opinion.

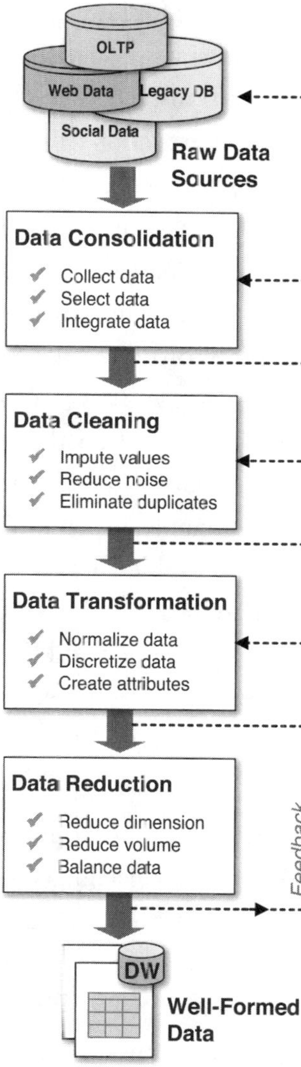

Figure 4.2 A Systematic Process for Data Preparation

In the third phase of data preprocessing, the data is transformed for better processing. For instance, in many cases, data is normalized between a certain minimum and maximum for all variables in order to mitigate the potential bias of one variable dominating other variables having smaller values; for example, the analyst would not want data with large numeric values, such as for household income, to dominate the number of dependents or years in service, which may potentially be more important. Another transformation that takes place at this point is discretization and/or aggregation. In some cases, the numeric variables are converted to categorical values (e.g., low, medium, high); in other cases, a nominal variable's unique value range is reduced to a smaller set, using concept hierarchies (e.g., instead of using the individual states with 50 different values, using several regions for a variable that shows location) in order to have a data set that is more amenable to computer processing. In yet other cases, one might choose to create new variables based on the existing ones in order to magnify the information found in a collection of variables in the data set. For instance, in an organ transplantation data set, an analyst might choose to use a single variable showing the blood type match (1: match, 0: no match) as opposed to using separate multinomial values for the blood types of the donor and the recipient. Such simplification may increase the information content while reducing the complexity of the relationships in the data.

The final phase of data preprocessing is data reduction. Data miners like to have large data sets, but too much data can be a problem. In the simplest sense, one can visualize the data commonly used in data mining projects as a flat file consisting of two dimensions: variables (the number of columns) and cases/records (the number of rows). In some cases (e.g., image processing, genome projects with complex microarray data), the number of variables can be rather large, and the analyst must reduce the number down to a manageable size. Because the variables are treated as different dimensions that describe the phenomenon from different perspectives, in data mining, this process is commonly called *dimensional reduction*. Even though there is not a single best way to accomplish this task, an analyst can use the findings from previously published literature, consult domain experts, run appropriate statistical tests (e.g., principal component analysis, independent component analysis), or, ideally, use a combination of these techniques to successfully reduce the dimensions in the data into a more manageable and relevant subset.

In addition to the complications related to multidimensionality, real-world data may also have issues with very large numbers of samples. That is, some data sets may include millions or billions of *records* (also called *samples* or *cases*). Even though computing power is increasing steadily, providing more horsepower for data analysis, processing such a large number of records may still not be feasible or practical; a single analysis may take a prohibitively long time to complete. In such cases, an analyst may need to analyze only a subset of the data. The underlying assumption of sampling is that the subset of the data will contain all relevant patterns of the complete data set. In a homogenous data set, such an assumption may hold well, but real-world data is hardly ever homogenous. The analyst should be extremely careful in selecting a subset of the data that reflects the essence of the complete data set and is not specific to a subgroup or subcategory. The data is usually sorted on some variable, and taking a section of the data from the top or bottom may lead to a biased data set on specific values of the indexed variable; therefore, one should always try to randomly select the records on the sample set. For skewed data, straightforward random sampling may not be sufficient, and stratified sampling—in which a proportional representation of different subgroups in the data is represented in the sample data set—may be required. In addition, it is a good practice to balance highly skewed data by either oversampling the less represented classes or undersampling the more represented classes. Research has shown that balanced data sets tends to produce better prediction models than unbalanced ones.

The process shown in Figure 4.2 is not a simple unidirectional, one-pass (top-to-bottom) effort. Rather, it is an iterative process that usually requires going backward (following the feedback loops) to make corrections and adjustments. The quality of outcomes (i.e., how well-formed the data ends up being) depends on the meticulous execution of this highly repetitive process between the later and earlier phases. The core activities in data preprocessing are summarized in Table 4.1, which maps the main tasks (along with their problem descriptions) to a representative list of methods and algorithms.

Table 4.1 Summary of Data Preprocessing Tasks and Potential Methods

Main Task	Subtasks	Popular Methods
Data consolidation	Access and collect the data	SQL queries, software agents, Web crawlers data access APIs
	Select and filter the data	Domain expertise, SQL queries, statistical tests
	Integrate and unify the data	SQL queries, domain expertise, ontology-driven data mapping
Data cleaning	Handle missing values in the data	Filling in the missing values (imputations) with most appropriate values (mean, median, min/max, mode, etc.); recoding of the missing values with a constant such as "ML"; removal of the record of the missing value; doing nothing
	Identify and reduce noise in the data	Identification of outliers in data with simple statistical techniques (e.g., averages and standard deviations) or with cluster analysis; removal of the outliers or smoothing by using binning, regression, or simple averages
	Find and eliminate erroneous data	Identification of erroneous values in data (other than outliers), such as odd values, inconsistent class labels, odd distributions; use of domain expertise to correct the values or remove the records that contain erroneous values
Data transformation	Normalize the data	Reduction of the range of values in each numerically valued variable to a standard range (e.g., 0 to 1, –1 to +1) by using a variety of normalization or scaling techniques

Main Task	Subtasks	Popular Methods
	Discretize or aggregate the data	If needed, conversion of the numeric variables into discrete representations using range- or frequency-based binning techniques; for categorical variables, reduction of the number of values via application of proper concept hierarchies
	Construct new attributes	Derivation of new and more informative variables from the existing ones, using a wide range of mathematical functions (as simple as addition and multiplication or as complex as a hybrid combination of log transformations)
Data reduction	Reduce number of attributes	Principal component analysis, independent component analysis, chi-square testing, correlation analysis, decision tree induction
	Reduce number of records	Random sampling, stratified sampling, expert knowledge-driven purposeful sampling
	Balance skewed data	Oversampling of the less represented or undersampling of the more represented class records

Data Mining Methods

A wide variety of data mining methods and algorithms are being developed and used by businesses and scientific organizations to shed light on a range of complex problems. These problems can be mapped to a few high-level data mining tasks. Depending on the need and the nature of the problem, the data mining task can be classified as prediction (either classification or regression), association, or clustering. Each of these data mining tasks can use a variety of data mining methods and algorithms. Figure 4.3 shows a graphical representation of these data mining tasks, methods, and algorithms. The following sections describe these broad categories of data mining tasks (i.e., prediction, association, and clustering) and the algorithms most commonly used with them (e.g., decision trees for prediction, Apriori algorithm for association, *k*-means for clustering). Chapter 5, "Data Mining Algorithms," explains several other data mining algorithms, including *k*-nearest neighbor, artificial neural networks, support vector machines, linear regression, logistic regression, and time-series forecasting. Most data mining software tools employ more than one technique (or algorithm) for each of these data mining tasks/methods.

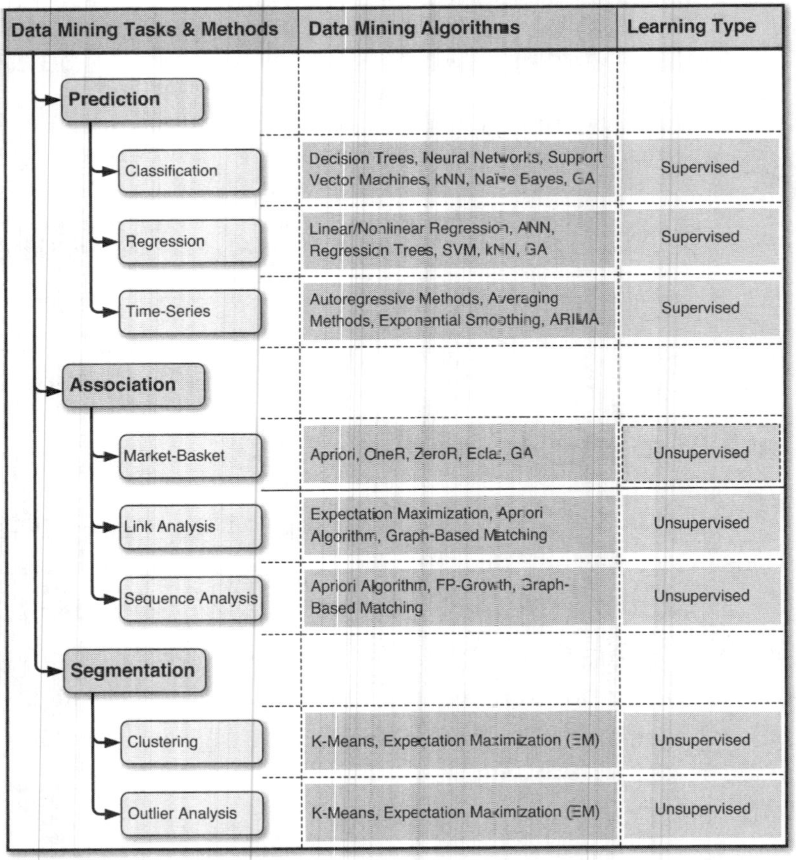

Figure 4.3 Data Mining Tasks/Methods and Representative Algorithms

Prediction

As discussed in Chapter 1, "Introduction to Analytics," prediction (or predictive analytics) is used to foresee the future and answer the question "What will happen?" Business planning and problem solving rely heavily on the estimation of the future occurrences. Predictive analytics (and underlying prediction modeling techniques) are the enablers of these estimates. Regardless of the modeling technique used to do the predictive modeling, the underlying assumption is that what we already know about the past (in terms of recorded transaction and events, in the form of historical data) is what we use to predict the future. We assume that the past will repeat itself in the future; if it doesn't, then there is no way of predicting the future.

Based on what we are predicting, the method of prediction can be labeled as *classification* (if we are predicting a discrete outcome), *regression* (if we are predicting a numerical value), or *time-series forecasting* (if we are predicting the future projection of a time-dependent value series). For instance, predicting whether a stock will close up or down at the end of the day is considered a classification, predicting the value difference of the stock at the end of the day is considered a regression, and predicting tomorrow's closing price of the stock based on the closing prices in the recent past is a time-series prediction. Even though they all are estimating the future value of the variable, the value type and the nature of the variable determine the classification of the prediction.

Classification

Classification is perhaps the most commonly used data mining method for real-world problems. As a popular member in the machine learning family of techniques, classification learns patterns from past data (a set of information—traits, variables, and features—on characteristics of the previously labeled items, objects, or events) in order to place new instances (with unknown labels) into their respective groups or classes. For example, one could use classification to predict whether the weather on a particular day will be "sunny," "rainy," or "cloudy." Popular classification tasks include credit approval (i.e., good or bad credit risk), store location (e.g., good/moderate/bad), target marketing (e.g., likely customer/no hope), fraud detection (i.e., yes/no), and telecommunication (e.g., likely/not likely to turn to another phone company). If what is being predicted is a class label (e.g., "sunny," "rainy," or "cloudy") the prediction problem is called a *classification*, whereas if it is a numeric value (e.g., temperature such as 68°F), the prediction problem is called a *regression*.

Even though clustering (another popular data mining method) can also be used to determine groups (or class memberships) of things, there is a significant difference between the two. Classification learns the function between the characteristics of things (i.e., independent variables) and their membership (i.e., output variable) through a supervised learning process where both types (input and output) of variables are presented to the algorithm; in clustering, the membership of the objects is learned through an unsupervised learning process where only the input variables are presented to the algorithm. Unlike classification, clustering does not have a supervising (or controlling) mechanism that enforces the learning process; instead, clustering algorithms use one or more heuristics (e.g., multidimensional distance measure) to discover natural groupings of objects.

The most common two-step methodology of classification-type prediction involves model development/training and model testing/deployment. In the model development phase, a collection of input data, including the actual class labels, is used. After a model has been trained, the model is tested against the holdout sample for accuracy assessment and eventually deployed for actual use, where it predicts classes of new data instances (where the class label is unknown).

Several factors are considered in assessing the model, including the following:

- **Classification accuracy.** This is the model's ability to correctly predict the class label of new or previously unseen data. Prediction accuracy is the most commonly used assessment factor for classification models. To compute this measure, actual class labels of a test data set are matched against the class labels predicted by the model. The accuracy can then be computed as the accuracy rate, which is the percentage of test data set samples correctly classified by the model. (More on this topic is provided later in the chapter.)
- **Speed.** Speed is the computational cost involved in generating and using the model, where faster is deemed to be better.
- **Robustness.** Robustness refers to a model's ability to make reasonably accurate predictions, given noisy data or data with missing and erroneous values.
- **Scalability.** Scalability is the ability to construct a prediction model efficiently, given a rather large amount of data.
- **Interpretability.** Interpretability is the level of understanding and insight provided by the model (e.g., how and/or what the model concludes on certain predictions).

How to Estimate the True Accuracy of Classification Models

In classification problems, the primary source for accuracy estimation is a *confusion matrix*, also called a *classification matrix* or a *contingency table*. Figure 4.4 shows a confusion matrix for a two-class classification problem. The number of instances recorded along the diagonal from the upper left to the lower right represents correctly made prediction decisions, and the number of instances recorded outside this diagonal represents errors (i.e., incorrectly made predictions).

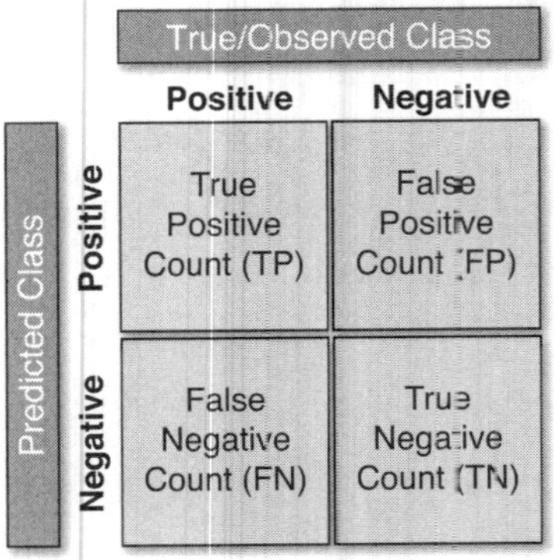

Figure 4.4 A Confusion Matrix for Binary Classification

When the classification problem is not binary, the confusion matrix gets bigger (it becomes an *n*-by-*n* square matrix for an output variable with *n* unique number of class values), and accuracy metrics become limited to per-class accuracy rates and the overall classifier accuracy. Table 4.2 lists some of the most commonly used accuracy metrics for classification models.

Table 4.2 Common Accuracy Metrics for Classification Models

Metric	Description
$True\ Positive\ Rate = \frac{TP}{TP + FN}$	The ratio of correctly classified positives divided by the total positive count (i.e., hit rate or recall)
$True\ Negative\ Rate = \frac{TN}{TN + FP}$	The ratio of correctly classified negatives divided by the total negative count (i.e., false-alarm rate)
$Accuracy = \frac{TP + TN}{TP + TN + FP + FN}$	The ratio of correctly classified instances (positives and negatives) divided by the total number of instances

Metric	Description
$Precision = \frac{TP}{TP + FP}$	The ratio of correctly classified positives divided by the sum of correctly classified positives and incorrectly classified positives
$Recall = \frac{TP}{TP + FN}$	The ratio of correctly classified positives divided by the sum of correctly classified positives and incorrectly classified negatives
$TrueClassificationRate_i = \frac{TrueClassification_i}{\sum FalseClassification_i}$	Classification accuracy for class *i* for a multiclass problem
$OverallAccuracy = \frac{\sum TrueClassification_i}{Number\ of\ Cases}$	Overall classification accuracy for a multiclass classification problem

Estimating the accuracy of a classification model (or classifier) induced by a supervised learning algorithm is important for two reasons. First, the estimation results can be used to assess the future prediction accuracy, which could imply the level of confidence in the classifier's output in the actual prediction system. Second, the estimation results can be used for choosing a classifier from a given set (identifying the "best" classification model among the many trained). The following sections describe some of the most popular estimation methodologies used for classification-type data mining models.

Simple Split

The simple split methodology (or holdout methodology or test sample estimation) partitions data into two mutually exclusive subsets called a training set and a test set (or holdout set). It is common to designate two-thirds of the data as the training set and the remaining one-third as the test set. The training set is used by the inducer (model builder), and the built classifier is then tested on the test set. An exception to this rule occurs when the classifier is an artificial neural network. In that case, the data is partitioned into three mutually exclusive subsets: training, validation, and testing. The validation set is used during model building to prevent overfitting. (For more on artificial neural networks, see Chapter 5.) Figure 4.5 shows the simple split methodology.

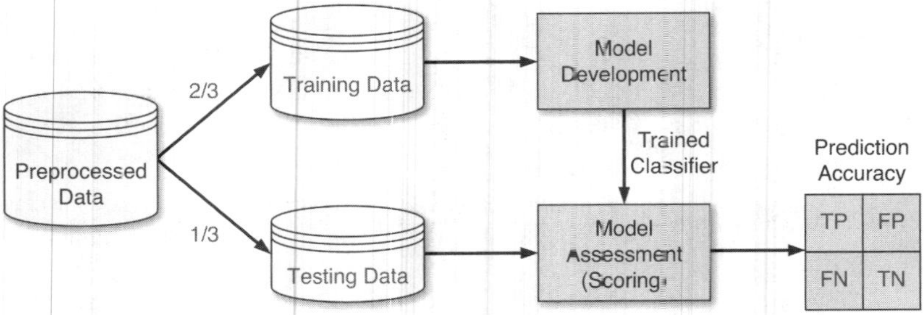

Figure 4.5 Simple Random Data Splitting for Training and Testing

The main criticism of this method is that it makes the assumption that the data in the two subsets is of the same kind (i.e., has exactly the same properties). Because this is a simple random partitioning, in most realistic data sets where the data is skewed on the classification variable, such an assumption may not hold true. In order to improve this situation, stratified sampling is suggested, where the strata become the output variable. Even though this is an improvement over the simple split methodology, it still has a bias associated with the single random partitioning.

k-Fold Cross-Validation

In order to minimize the bias associated with the random sampling of the training and holdout data samples in comparing the predictive accuracy of two or more methods, an analyst can use a methodology called *k*-fold cross-validation. In this methodology, also called rotation estimation, the complete data set is randomly split into *k* mutually exclusive subsets of approximately equal size. The classification model is trained and tested *k* times. Each time, it is trained on all but one fold and then tested on the remaining single fold. The cross-validation estimate of the overall accuracy of a model is calculated by simply averaging the k individual accuracy measures. This is the formula for this aggregation:

$$CrossValidationAccuracy = \frac{1}{k}\sum_{i=1}^{k} Accuracy_i \quad (k\text{: Number of folds})$$

A graphical illustration of k-fold cross-validation, where the k is set to 10 (i.e., 10-fold cross-validation) is shown in Figure 4.6.

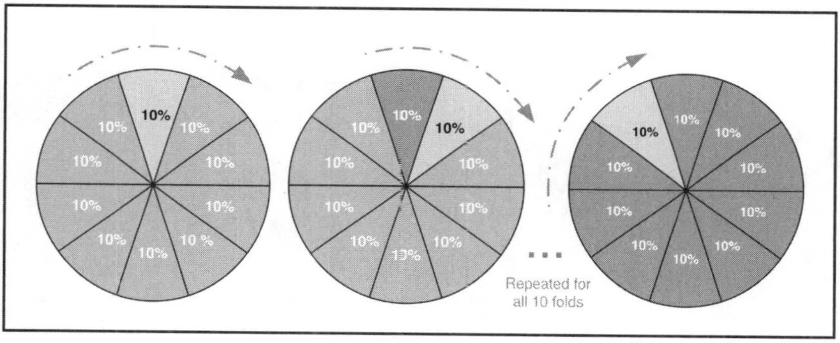

Figure 4.6 A Graphical Illustration of 10-fold Cross-Validation Methodology

Other Classification Assessment Methodologies

Although not as common as the ones describe above, there are other assessment techniques used for classification type problems. Here are some of the most popular ones:

- **Leave-one-out.** The leave-one-out method is similar to a k-fold cross-validation where k takes the value 1; that is, every data point is used for testing once on as many models developed as there are data points. This is a time-consuming methodology, but it can be a viable option for small data sets.
- **Bootstrapping.** With bootstrapping, a fixed number of instances from the original data is sampled (with replacement) for training, and the rest of the data set is used for testing. This process is repeated as many times as desired.

- **Jackknifing.** This methodology is similar to the leave-one-out methodology; with it, accuracy is calculated by leaving one sample out at each iteration of the estimation process.
- **Area under the ROC curve.** The area under the ROC curve is a graphical assessment technique in which the true positive rate is plotted on the *y*-axis, and the false positive rate is plotted on the *x*-axis. The area under the ROC curve determines the accuracy measure of a classifier: A value of 1 indicates a perfect classifier, whereas 0.5 indicates no better than random chance, though in reality, the values would range between the two extreme cases. For example, in Figure 4.7, A has a better classification performance than B, while C is not any better than the random chance involved in flipping a coin.

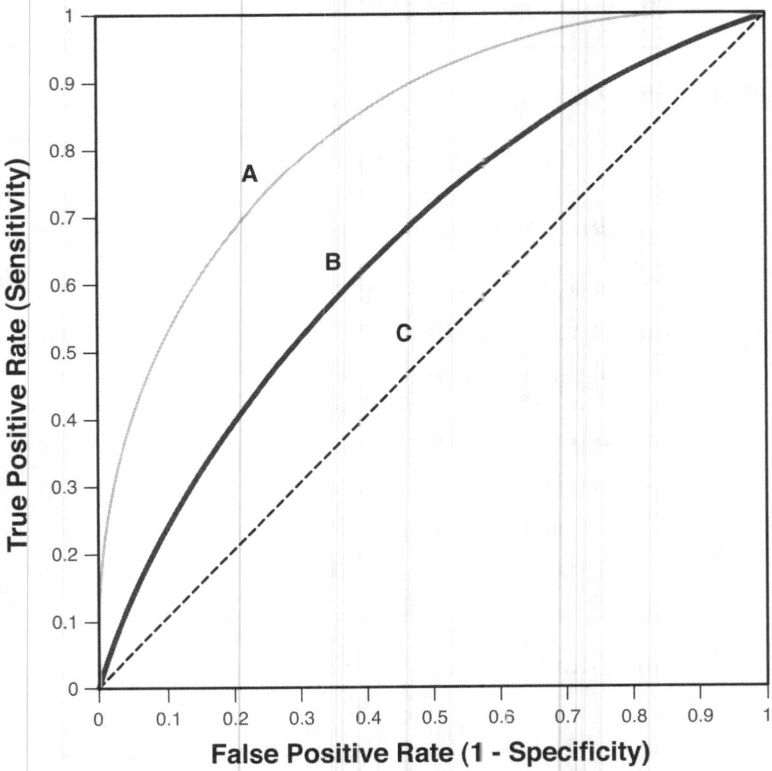

Figure 4.7 An Example of Area Under the ROC Curve

Classification Methods

Many methods and algorithms are used for classification modeling, including the following:

- **Decision tree analysis.** Decision tree analysis (a machine learning technique) is arguably the most popular classification technique in the data mining arena. A detailed description of this technique is given in the following section.
- **Statistical analysis.** Statistical techniques were the primary classification algorithm for many years, until the emergence of machine learning techniques. Statistical classification techniques include *logistic regression* and *discriminant analysis*, both of which make the assumptions that the relationships between the input and output variables are linear in nature, the data is normally distributed, and the variables are not correlated and are independent of each other. The questionable nature of these assumptions has led to the shift toward machine learning techniques.
- **Neural networks.** Using neural networks is among the most popular machine learning techniques that can be used for classification-type problems. A detailed description of this technique is presented in Chapter 5.
- **Support vector machines.** Along with neural networks, support vector machines are also increasing in popularity as powerful classification algorithms. A detailed description of this technique is presented in Chapter 5.
- **Nearest-neighbor algorithm.** This deceptively simple yet highly efficient algorithm uses similarity as the basis for its classification method. A detailed description of this technique is also presented in Chapter 5.
- **Case-based reasoning.** Conceptually similar to the nearest-neighbor algorithm, this approach uses historical cases

to recognize commonalities in order to assign a new case to the most probable category.

- **Bayesian classifiers.** This approach uses probability theory to build classification models based on the past occurrences that are capable of placing a new instance into a most probable class (or category).
- **Genetic algorithms.** Genetic algorithms use the analogy of natural evolution to build directed-search-based mechanisms to classify data samples.
- **Rough sets.** This method takes into account the partial membership of class labels to predefined categories in building models (collection of rules) for classification problems.

Complete coverage of all these classification techniques is beyond the scope of this book, so here we describe the most popular one, decision trees, and we cover a few of the other ones in Chapter 5.

Decision Trees

Before describing the details of decision trees, we need to discuss some simple terminology. First, decision trees include many input variables that may have an impact on the classification of different patterns. These input variables are usually called *attributes*. For example, if we were to build a model to classify loan risks on the basis of just two characteristics—income and credit rating—these two characteristics would be the attributes, and the resulting output would be the class label (e.g., low, medium, or high risk). Second, a tree consists of branches and nodes. A branch represents the outcome of a test to classify a pattern (on the basis of a test), using one of the attributes. A leaf node at the end represents the final class choice for a pattern (a chain of branches from the root node to the leaf node, which can be represented as a complex if–then statement).

The basic idea behind a decision tree is that it recursively divides a training set until each division consists entirely or primarily of examples from one class. Each nonleaf node of the tree contains a split point, which is a test on one or more attributes and determines how the data is to be divided further. Decision tree algorithms, in general, build an initial tree from the training data such that each leaf node is pure, and they then prune the tree to increase its generalization and, hence, the prediction accuracy of the test data.

In the growth phase, the tree is built by recursively dividing the data until each division is either pure (i.e., contains members of the same class) or relatively small. The basic idea is to ask questions whose answers provide the most information, as is done in the game "Twenty Questions."

The split used to partition the data depends on the type of attribute used in the split. For a continuous attribute A, splits are of the form value(A) *x*, where *x* is some "optimal" split value of A. For example, the split based on income could be "Income > 50000." For the categorical attribute A, splits are of the form value(A) belongs to *x*, where *x* is a subset of A. For example, the split could be on the basis of gender: male versus female.

A general algorithm for building a decision tree is as follows:

1. Create a root node and assign all the training data to it.
2. Select the best splitting attribute.
3. Add a branch to the root node for each value of the split. Split the data into mutually exclusive (i.e., non-overlapping) subsets along the lines of the specific split of the branches.
4. Repeat steps 2 and 3 for each leaf node until the stopping criteria is reached (e.g., the node is dominated by a single class label).

Many different algorithms have been proposed for creating decision trees. These algorithms differ primarily in terms of the way in which they determine the splitting attribute (and its split values), the order of splitting the attributes (splitting the same attribute only once or many times), the number of splits at each node (binary versus ternary), the stopping criteria, and the pruning of the tree (pre- versus postpruning). Some of the most well-known algorithms are ID3 (followed by C4.5 and C5 as the improved versions of ID3) from machine learning, classification and regression trees (CART) from statistics, and the chi-squared automatic interaction detector (CHAID) from pattern recognition.

When building a decision tree, the goal at each node is to determine the attribute and the split point of that attribute that best divides the training records in order to purify the class representation at that node. To evaluate the goodness of the split, some splitting indices have been proposed. Two of the most common ones are the Gini index and information gain. The Gini index is used in the CART and SPRINT (Scalable Parallelizable Induction of Decision Trees) algorithms. Versions of information gain are used in ID3 (and its newer versions, C4.5 and C5).

The Gini index has been used in economics to measure the diversity of a population. The same concept can be used to determine the purity of a specific class as a result of a decision to branch along a particular attribute or variable. The best split is the one that increases the purity of the sets resulting from a proposed split.

Information gain is the splitting mechanism used in ID3, which is perhaps the most widely known decision tree algorithm. It was developed by Ross Quinlan in 1986, and since then Quinlan has evolved this algorithm into the C4.5 and C5 algorithms. The basic idea behind ID3 and its variants is to use a concept called *entropy* in place of the Gini index. Entropy measures the extent of uncertainty or randomness in a data set. If all the data in a subset belongs to just one class, there is no uncertainty or randomness in that data set, so the entropy is zero. The objective of this approach is to build subtrees so that the entropy of each final subset is zero or close to zero.

Cluster Analysis for Data Mining

Cluster analysis is an essential data mining method for classifying items, events, or concepts into common groupings called *clusters*. The method is commonly used in biology, medicine, genetics, social network analysis, anthropology, archaeology, astronomy, character recognition, and even MIS development. As the popularity of data mining has increased, the underlying techniques have been applied to business, especially to marketing. Cluster analysis has been used extensively for fraud detection (both credit card and ecommerce fraud) and market segmentation of customers in contemporary CRM systems. More applications in business continue to be developed as the strength of cluster analysis is recognized and used.

Cluster analysis is an exploratory data analysis tool for solving classification problems. The objective is to sort cases (e.g., people, things, events) into groups, or clusters, so that the degree of association is strong among members of the same cluster and weak among members of different clusters. Each cluster describes the class to which its members belong. An obvious one-dimensional example of cluster analysis is to establish score ranges into which to assign class grades for a college class. This is similar to the cluster analysis problem that the U.S. Treasury faced when establishing new tax brackets in the 1980s. A fictional example of clustering occurs in J. K. Rowling's Harry Potter books. The Sorting Hat determines to which House (e.g., dormitory) to assign first-year students at the Hogwarts School. Another example involves determining how to seat guests at a wedding. As far as data mining goes, the importance of cluster analysis is that it may reveal associations and structures in data that were not previously apparent but are sensible and useful once found.

Cluster analysis results may be used to:

- Identify a classification scheme (e.g., types of customers)
- Suggest statistical models to describe populations

- Indicate rules for assigning new cases to classes for identification, targeting, and diagnostic purposes
- Provide measures of definition, size, and change in what were previously broad concepts
- Find typical cases to label and represent classes
- Decrease the size and complexity of the problem space for other data mining methods
- Identify outliers in a specific domain (e.g., rare-event detection)

How to Determine the Number of Clusters

Clustering algorithms usually require specifying the number of clusters to find. If this number is not known from prior knowledge, it should be chosen in some way. Unfortunately, there is not an optimal way of calculating what this number is supposed to be. Therefore, several different heuristic methods have been proposed. The following are among the most commonly referenced ones:

- Look at the percentage of variance, expressed as a function of the number of clusters; that is, choose a number of clusters so that adding another cluster would not give much better modeling of the data. Specifically, in graphing the percentage of variance explained by the clusters, there is a point at which the marginal gain will drop (giving an angle in the graph), and this indicates the number of clusters to be chosen.
- Set the number of clusters to $(n/2)^{1/2}$, where n is the number of data points.
- Use the Akaike information criterion (AIC), which is a measure of the goodness of fit (based on the concept of entropy) to determine the number of clusters.
- Use the Bayesian information criterion (BIC), which is a model-selection criterion (based on maximum likelihood estimation) to determine the number of clusters.

Clustering Methods

Cluster analysis involves grouping a set of records (samples, objects) in such a way that objects in the same group (called a *cluster*) are more similar to each other than to those in other groups (or clusters). Most cluster analysis methods involve the use of a distance measure to calculate the closeness between pairs of items. Popular distance measures include Euclidian distance (the ordinary distance between two points that one would measure with a ruler) and Manhattan distance (also called rectilinear distance, or taxicab distance, between two points). There are a number of cluster analysis techniques, each having a slightly different way of identifying natural grouping of objects or records. Figure 4.8 shows a simple taxonomy of clustering techniques.

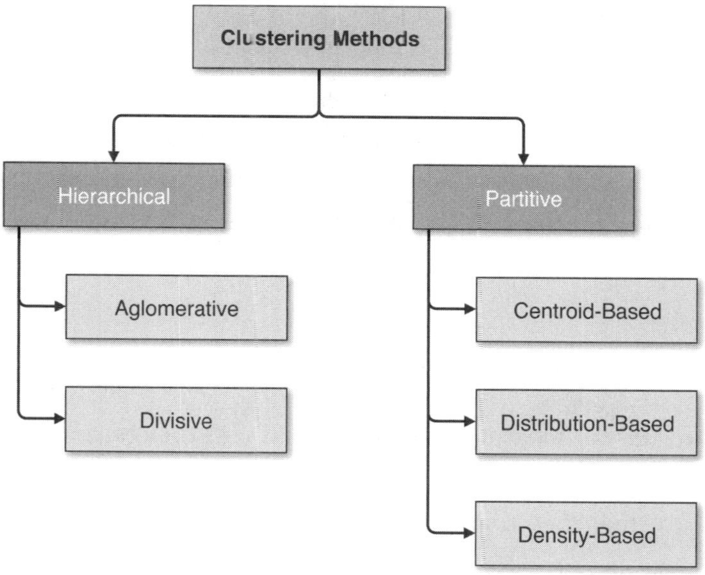

Figure 4.8 A Simple Taxonomy of Clustering Methods

As shown in Figure 4.8, at the highest level, the clustering methods can be classified as hierarchical and partitive. *Hierarchical clustering methods* are based on the basic idea that nearby objects are more related to each other than are those that are farther away. Therefore, a cluster can be defined by the maximum distance needed to connect all related objects. At different distances, different clusters would form, which can be represented using a dendrogram, which explains where the name *hierarchical clustering* comes from. In hierarchical clustering, the clusters can be identified either bottom-up (i.e., agglomerative—starting from the bottom, where each object is a cluster of its own and then adding similar objects to create more general clusters) or top-down (i.e., divisive—starting from the top cluster and then splitting the objects into more specific clusters).

Partitive clustering methods, on the other hand, are based on the idea that there are natural groups of objects in a multidimensional space. These methods follow the process of starting from an arbitrary number of clusters, assigning objects to most probable clusters, and then updating the number of clusters and evolving the assignment of objects to clusters until a stable cluster space is reached. The specific algorithms under partitive clustering use different ways to identify the clusters. Since there is not a single best method or algorithm for clustering, an analyst working with a clustering task should try several competing algorithms and assess their goodness of fit and identify the best one.

Assessment of a Clustering Technique

How do we measure the goodness of clusters created by a clustering method? How do we know if they are valid? How do we know to what extent they represent the natural grouping of the data? We need measures to compare how well different clustering algorithms perform on a given set of data. We can then choose the "best" one to use for a classification-type data mining problem. The measures for assessing the goodness of clustering methods are often grouped into two categories: internal assessment and external assessment.

Internal Assessment

Internal assessment methods usually assign the best assessment score to the method (or algorithm) that produces clusters with high similarity within a cluster and low similarity between clusters. The internal evaluation measures are best suited to situations where one algorithm performs better than another, but this may not imply that one algorithm produces more "valid" results than another. Validity is different from comparative assessment because it suggests that the identified cluster structure truly exists in the data set. This assessment methods measures intraclass and interclass similarities using some type of a distance measure in a multidimensional feature space.

One drawback of using internal criteria in cluster evaluation is that high scores on an internal measure do not necessarily result in effective information-retrieval applications. In addition, this evaluation is biased toward algorithms that use the same cluster model. For example, *k*-means clustering naturally optimizes object distances, and a distance-based internal criterion will likely overrate the resulting clustering.

External Assessment

Whereas in internal evaluation, clustering results are evaluated based on similarity measures, in external evaluation, the cluster results are measured based on data (with known class labels) that was not used for clustering. This assessment consists of a set of preclassified records, with known classes or class labels created by domain experts. Thus, the comparison sets can be thought of as a gold standard for evaluation. These types of evaluation methods measure how close the clustering is to the predetermined labeled classes. As logical as it may sound, widespread practicality of this method is questionable. Finding prelabeled classes for real-world data sets is difficult and may even be impossible. Also, because the true classes contain some type of natural structure, the attributes/variables presented to the clustering algorithm may not be complete enough to allow separation of true clusters.

k-Means Clustering Algorithm

The *k*-means algorithm (where *k* is the predetermined number of clusters) is arguably the most referenced clustering algorithm. It has its roots in traditional statistical analysis. As part of the family of centroid-based partitive clustering methods, the *k*-means algorithm, as its name implies, assigns each data point (e.g., customer, event, object) to the cluster whose center (also called *centroid*) is the nearest. The center is calculated as the average of all the points in the cluster; that is, its coordinates are the arithmetic mean for each dimension separately over all the points in the cluster. The algorithm steps are listed here and shown graphically in Figure 4.9:

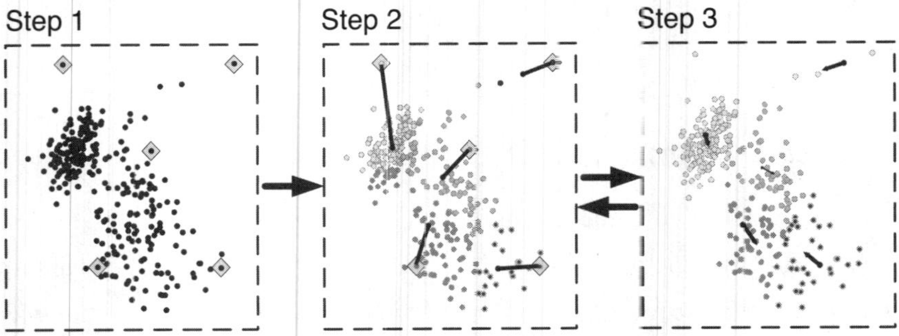

Figure 4.9 Steps in the k-Means Clustering Algorithm

Initialization step: Choose the number of clusters (i.e., the value of *k*).

1. Randomly generate *k* random points as initial cluster centers.
2. Assign each point to the nearest cluster center.
3. Recompute the new cluster centers.

Repetition step: Repeat steps 2 and 3 until some convergence criterion is met (usually that the assignment of points to clusters becomes stable).

Association

Association—also known as association rule mining, or affinity analysis, or market-basket analysis—is a popular data mining method that is commonly used as an example to explain what data mining is and what it can do. You might have heard the famous (or infamous, depending on how you look at it) relationship discovered between the sales of beer and diapers at grocery stores. As the story goes, a large supermarket chain (maybe Wal-Mart, maybe not; there is no consensus on which supermarket chain it was) did an analysis of customers' buying habits and found a statistically significant correlation between purchases of beer and purchases of diapers. It was theorized that the reason for this was that fathers (presumably young men) were stopping off at the supermarket to buy diapers for their babies (especially on Thursdays), and since they could no longer go down to the sports bar as often, they would buy beer as well. As a result of this finding, the supermarket chain is alleged to have placed the diapers next to the beer, resulting in increased sales of both.

In essence, association rule mining aims to find interesting relationships (affinities) between variables (items) in large databases. Because of its successful application to retail business problems, it is commonly called market-basket analysis. The main idea in market-basket analysis is to identify strong relationships among different products (or services) that are usually purchased together (i.e., show up in the same basket together, either a physical basket at a grocery store or a virtual basket at an ecommerce Website). For example, 65% of those who buy comprehensive automobile insurance also buy health insurance; 80% of those who buy books online also buy music online;

60% of those who have high blood pressure and are overweight have high cholesterol; 70% of the customers who buy laptop computer and virus protection software also buy extended service plans.

The input to market-basket analysis is simple point-of-sale transaction data, where a number of products and/or services purchased together (just like the content of a purchase receipt) are tabulated under a single transaction instance. The outcome of the analysis is invaluable information that can be used to better understand customer-purchase behavior in order to maximize the profit from business transactions. A business can take advantage of such knowledge by (1) putting the items next to each other to make it more convenient for the customers to pick them up together and not forget to buy one when buying the others (increasing sales volume); (2) promoting the items as a package (e.g., not putting one on sale if the other is on sale); and (3) placing the items apart from each other so that the customer has to walk the aisles to search for them and by doing so will potentially see and buy other items.

Applications of market-basket analysis include cross-marketing, cross-selling, store design, catalog design, ecommerce site design, optimization of online advertising, product pricing, and sales/promotion configuration. In essence, market-basket analysis helps businesses infer customer needs and preferences from their purchase patterns. Outside the business realm, association rules are successfully used to discover relationships between symptoms and illnesses, diagnosis and patient characteristics and treatments (which can be used in medical decision support systems [DSSs]), and genes and their functions (which can be used in genomics projects), among others. Here are a few common areas and uses for association rule mining:

- **Sales transactions.** Having knowledge of combinations of retail products purchased together can be used to improve product placement on sales floor (placing products that go together in close proximity) and promotional pricing of products (not having promotions on two products that are often purchased together).

- **Credit card transactions.** Having knowledge of items purchased with a credit card provides insight into other products the customer is likely to purchase or fraudulent use of a credit card number.
- **Banking services.** The sequential patterns of services used by customers (e.g., checking account followed by saving account) can be used to identify other services they may be interested in (e.g., investment account).
- **Insurance service products.** Bundles of insurance products bought by customers (e.g., car insurance followed by a home insurance) can be used to propose additional insurance products (e.g., life insurance); on the other hand, unusual combinations of insurance claims can be a sign of fraud.
- **Telecommunication services.** Commonly purchased group of options (e.g., call waiting, caller ID, three-way calling) help better structure product bundles to maximize revenue; this is also applicable to multichannel telecom providers with phone, television, and Internet service offerings.
- **Medical records.** Certain combinations of conditions can indicate increased risk of various complications; or certain treatment procedures at certain medical facilities can be tied to certain types of infection.

A good question to ask with respect to the patterns/relationships that association rule mining can discover is "Are all association rules interesting and useful?" In order to answer such a question, association rule mining uses three common metrics: support, confidence, and lift. Before defining these terms, let us look at what an association rule looks like. For example, in this rule, *X* (products and/or service; called the left-hand side [LHS], or antecedent) is associated with Y (products and/or service; called the right-hand side [RHS], or consequent):

$$X \Rightarrow Y \; [S(\%), C(\%)]$$

S is the support, and C is the confidence for this particular rule. Here are the simple formulas for support, confidence, and lift:

$$Support = Support(X \Rightarrow Y) = \frac{number\ of\ baskets\ that\ contain\ both\ X\ and\ Y}{total\ number\ of\ baskets}$$

$$Confidence = Confidence(X \Rightarrow Y) = \frac{Supp(X \Rightarrow Y)}{Supp(X)}$$

$$Lift(X \Rightarrow Y) = \frac{Conf(X \Rightarrow Y)}{Expected\ Conf(X \Rightarrow Y)} = \frac{\frac{S(X \Rightarrow Y)}{S(X)}}{\frac{S(X)*S(Y)}{S(X)}} = \frac{S(X \Rightarrow Y)}{S(X)*S(Y)}$$

The support (S) of a collection of products and/or services is the measure of how often those products and/or services (i.e., LHS + RHS = laptop computer, antivirus software, and extended service plan) appear together in the same transaction—that is, the proportion of transactions in the data set that contain all the products and/or services mentioned in a specific rule. In this example, 30% of all transactions in the hypothetical store database had all three products present on a single sales ticket. The confidence of a rule is the measure of how often the products and/or services on the RHS (consequent) go together with the products and/or services on the LHS (antecedent)—that is, the proportion of transactions that include LHS while also including the RHS. In other words, it is the conditional probability of finding the RHS of the rule present in transactions where the LHS of the rule already exists. The lift value of an association rule is the ratio of the confidence of the rule and the expected confidence of the rule. The expected confidence of a rule is defined as the product of the support values of the LHS and the RHS divided by the support of the LHS.

Several algorithms are available for discovering association rules. Some well-known algorithms include Apriori, Eclat, and FP-Growth. These algorithms do only half the job, which is to identify the frequent item sets in the database. Once the frequent item sets are identified, they need to be converted into rules with antecedent and consequent parts. Determination of the rules from frequent item sets is a straightforward matching process, but the process may be time-consuming with large transaction databases. Even though there can be many items in each section of the rule, in practice the consequent part usually contains a single item. The following section explains one of the most popular algorithms for identification of frequent item sets.

Apriori Algorithm

The Apriori algorithm is the algorithm most commonly used to discover association rules. Given a set of item sets (e.g., sets of retail transactions, each listing individual items purchased), the algorithm attempts to find subsets that are common to at least a minimum number of the item sets (i.e., comply with a minimum support). The Apriori algorithm uses a bottom-up approach, where frequent subsets are extended one item at a time (a method known as *candidate generation*, whereby the size of frequent subsets increases from one-item subsets to two-item subsets, then three-item subsets, etc.), and groups of candidates at each level are tested against the data for minimum support. The algorithm terminates when no further successful extensions are found.

As an illustrative example, consider the following. A grocery store tracks sales transactions by SKU (stock-keeping unit) and thus knows which items are typically purchased together. The database of transactions, along with the subsequent steps in identifying the frequent item sets, is shown in Figure 4.10. Each SKU in the transaction database corresponds to a product, such as "1 = butter," "2 = bread," "3 = water," and so on. The first step in the Apriori algorithm is to count up the frequencies (i.e., the supports) of each item (one-item item sets). For this overly simplified example, let us set the minimum support to 3 (or 50%; meaning an item set is considered to be a frequent item set if it shows up in at least 3 out of 6 transactions in the database). Because all the one-item item sets have at least 3 in the support column, they are all considered frequent item sets. However, if any of the one-item item sets were not frequent, they would not have been included as possible members of possible two-item pairs. In this way, Apriori prunes the tree of all possible item sets. As Figure 4.10 shows, using one-item item sets, all possible two-item item sets are generated, and the transaction database is used to calculate their support values. Because the two-item item set {1, 3} has a support less than 3, it should not be included in the frequent item sets that will be used to generate the next-level item sets (three-item item sets). The algorithm seems deceptively simple, but only for small data sets. In much larger data sets, especially those with huge numbers of items present in low quantities and small numbers of items present in big quantities, the search and calculation become computationally intensive.

Raw Transaction Data

Transaction No.	SKUs (Item No.)
1001234	1, 2, 3, 4
1001235	2, 3, 4
1001236	2, 3
1001237	1, 2, 4
1001238	1, 2, 3, 4
1001239	2, 4

One-Item Item Sets

Item Set (SKUs)	Support
1	3
2	6
3	4
4	5

Two-Item Item Sets

Item Set (SKUs)	Support
1, 2	3
1, 3	2
1, 4	3
2, 3	4
2, 4	5
3, 4	3

Three-Item Item Sets

Item Set (SKUs)	Support
1, 2, 4	3
2, 3, 4	3

Figure 4.10 The Process of Identifying Frequent Item Sets with the Apriori Algorithm

Data Mining Misconceptions and Realities

Data mining is a powerful analytical tool that enables business executives to advance from describing the nature of the past to predicting the future. It helps marketers find patterns that unlock the mysteries of customer behavior. The results of data mining can be used to increase revenue, reduce expenses, identify fraud, and locate business opportunities, offering a whole new realm of competitive advantage. As an evolving and maturing field, data mining is often associated with a number of misconceptions, including the ones listed in Table 4.3 (Zaima, 2003).

Table 4.3 A Few Misconceptions and Realities About Data Mining

Misconception	Reality
Data mining provides instant, crystal ball–like predictions.	Data mining is a multistep process that requires deliberate, proactive design and use.
Data mining is not yet viable for business applications.	The current state-of-the-art is ready to go for almost any business.
Data mining requires a separate, dedicated database.	Because of advances in database technology, a dedicated database is not required, even though it may be desirable.
Only those with advanced degrees can do data mining.	Web-based tools enable managers of all educational levels to do data mining.
Data mining is only for large firms that have lots of customer data.	If its data accurately reflects the business or its customers, a company can use data mining.

Data mining visionaries have gained enormous competitive advantage by understanding that these misconceptions are, in fact, misconceptions.

The following 10 data mining mistakes are often made in practice, and you should try to avoid them:

1. Selecting the wrong problem for data mining.
2. Ignoring what your sponsor thinks data mining is and what it really can and cannot do.

3. Leaving insufficient time for data preparation. This stage takes more effort than is generally understood.
4. Looking only at aggregated results and not at individual records. IBM's DB2 IMS can highlight individual records of interest.
5. Being sloppy about keeping track of the data mining procedure and results.
6. Ignoring suspicious findings and quickly moving on.
7. Running mining algorithms repeatedly and blindly. It is important to think hard about the next stage of data analysis. Data mining is a very hands-on activity.
8. Believing everything you are told about the data.
9. Believing everything you are told about your own data mining analysis.
10. Measuring your results differently from the way your sponsor measures them.

Application Example: Predicting NCAA Bowl Game Results

Predicting the outcome of a college football game (or any sports game, for that matter) is an interesting and challenging problem. Challenge-seeking researchers from both academics and industry have spent a great deal of effort on forecasting the outcomes of sporting events. Large quantities of historic data exist in different media outlets (and are often publicly available) regarding the structure and outcomes of sporting events, in the form of a variety of numerically or symbolically represented factors that are assumed to contribute to those outcomes. However, despite the large quantity of studies in sports (more than 43,000 hits on digital literature databases), a small percentage of papers have focused exclusively on the characteristics of sports forecasts. Instead, many papers have been written about the efficiency of sports markets. Since most previous betting-markets studies have been concerned with economic efficiency (Van Bruggen et al., 2010), they have not evaluated th [illegible] h those events. As [illegible] able amount of in [illegible] g process from stu [illegible] c efficiency (Stekler [illegible]

The end-of-seaso [illegible] s both financially (b [illegible] enue) and reputati [illegible] regarded high sch [illegible] that are selected to [illegible] size of which depe [illegible] prestigious and have higher payouts for the two teams. Therefore, securing an invitation to a bowl game is a big goal of any division I-A college football program. The bowl game decision makers are given the authority to select and invite bowl-eligible (a team that has six wins against its Division I-A opponents in that season) successful teams (as per the ratings and rankings) that will play in an exciting and competitive game, attract fans of both schools, and keep the remaining fans tuned in via a variety of media outlets for advertising.

In a recent data mining study, Delen et al. (2012) used eight years of bowl game data along with three popular data mining techniques (i.e., decision trees, neural networks, and support vector machines) to predict both the classification type outcome of a game (win versus loss) and the regression type outcome (i.e., projected point difference between the scores of the two opponents). What follows is a brief description of their study.

Methodology

In this research, Delen and his colleagues followed a popular data mining methodology called CRISP-DM (Cross-Industry Standard Process for Data Mining), which is a six-step process: (1) understand the domain and develop the goals for the study; (2) identify, access, and understand the relevant data sources; (3) preprocess, clean, and transform the relevant data; (4) develop models using comparable analytical techniques; (5) evaluate and assess the validity and utility of the models against each other and against the goals of the study; and (6) deploy the models for use in decision-making processes. This popular methodology, which is covered in detailed in Chapter 3, "The Data Mining Process," provided Delen et al. with a systematic and structured way to conduct the underlying data mining study and, hence, improved the likelihood of obtaining accurate and reliable results.

To objectively assess the predictive power of the different model types, Delen et al. used a cross-validation methodology, which is a popular statistical technique that is often used in data mining to compare multiple models in terms of their predictive accuracies. Traditional cross-validation methodology splits the data into two mutually exclusive subsets, training and testing (or three, in the case of neural networks that include validation sets). This type of single random split may lead to non-homogeneous representation in the subsets, especially in a small data set. In order to minimize the bias associated with the random sampling of the training and testing data samples, Delen et al. used *k*-fold cross-validation (discussed earlier in this chapter). Figure 4.11 provides a graphical representation of the methodology the researchers used.

Data

The sample data for this study was collected from a variety of sports databases available on the Web, including ESPN.com, Covers.com, ncaa.org, and rauzulusstreet.com. The data set included 244 bowl games, representing a complete set of eight seasons of college football bowl games played between 2002 and 2009. It also included an out-of-sample data set (2010–2011 bowl games) for additional validation purposes. Exercising one of the popular data mining rules-of-thumb, the authors included as much relevant information in the model as possible. Therefore, after an in-depth variable identification and collection process, they ended up with a data set that included 36 variables. The first 6 variables were the identifying variables: name and year of the bowl game, home and away team names, and their athletic conferences (see variables 1–6 in Table

4.4). The next 28 were input variables, which included variables delineating a team's seasonal statistics on offense and defense, game outcomes, team composition characteristics, athletic conference characteristics, and how they fared against the odds (see variables 7–34 in Table 4.4). The last 2 variables were the output variables: ScoreDiff (i.e., the score difference between the home team and the away team represented with an integer number) and WinLoss (i.e., whether the home team won or lost the bowl game, represented with a nominal label).

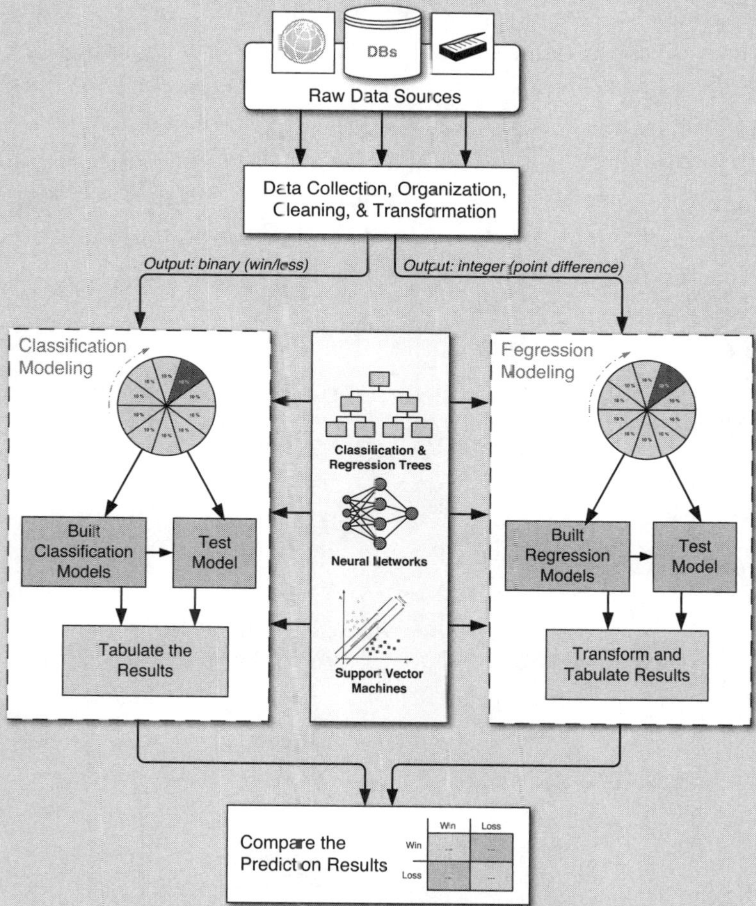

Figure 4.11 Graphical Illustration of the Methodology Employed in the Study

In the formulation of the data set, each row (a.k.a. tuple, case, sample, example) represented a bowl game, and each column stood for a variable (i.e., identifier/input or output type). In order to represent the game-related comparative characteristics of the two opponent teams, in the input variables, the authors calculated and used the differences between the measures of the home and away teams. All these variable values are calculated from the home team's perspective. For instance, the variable PPG (average number of points a team scored per game) represents the difference between the home team's PPG and the away team's PPG. The output variables represent whether the home team is expected to win or lose the bowl game. That is, if the ScoreDiff variable takes a positive integer number, then the home team is expected to win the game by that margin; if the ScoreDiff variable takes a negative integer number, then the home team is expected to lose the game by that margin. In the case of WinLoss, the value of the output variable is a binary label, Win or Loss, indicating the outcome of the game for the home team.

Table 4.4 Description of the Variables Used in the Study

No	Cat	Variable Name	Description
1	ID	YEAR	Year of the bowl game
2	ID	BOWL GAME	Name of the bowl game
3	ID	HOMETEAM	Home team (as listed by the bowl organizers)
4	ID	AWAYTEAM	Away team (as listed by the bowl organizers)
5	ID	HOMECONFERENCE	Conference of the home team
6	ID	AWAYCONFERENCE	Conference of the away team
7	I1	DEFPTPGM	Defensive points per game
8	I1	DEFRYDPGM	Defensive rush yards per game
9	I1	DEFYDPGM	Defensive yards per game
10	I1	PPG	Average number of points a given team scored per game
11	I1	PYDPGM	Average total pass yards per game

No	Cat	Variable Name	Description
12	I1	RYDPGM	Team's average total rush yards per game
13	I1	YRDPGM	Average total offensive yards per game
14	I2	HMWIN%	Home winning percentage
15	I2	LAST7	How many games the team won out of their past 7 games
16	I2	MARGOVIC	Average margin of victory
17	I2	NCTW	Non-conference team winning percentage
18	I2	PREVAPP	Whether the team appeared in a bowl game previous year
19	I2	RDWIN%	Road winning percentage
20	I2	SEASTW	Winning percentage for the year
21	I2	TOP25	Winning percentage against AP top 25 teams for the year
22	I3	TSOS	Strength of schedule for the year
23	I3	FR%	Percentage of games played by freshmen class players for the year
24	I3	SO%	Percentage of games played by sophomore class players for the year
25	I3	JR%	Percentage of games played by junior class players for the year
26	I3	SR%	Percentage of games played by senior class players for the year
27	I4	SEASOvUn%	Percentage of times a team went over the O/U* in the current season
28	I4	ATSCOV%	Against the spread cover percentage of the team in previous bowl games
39	I4	UNDER%	Percentage of times a team went under in previous bowl games
30	I4	OVER%	Percentage of times a team went over in previous bowl games
31	I4	SEASATS%	Percentage of covering against the spread for the current season

No	Cat	Variable Name	Description
32	I5	CONCH	Did the team win their respective conference championship game
33	I5	CONFSOS	Conference strength of schedule
34	I5	CONFWIN%	Conference winning percentage
35	O1	ScoreDiff**	Score difference (HomeTeamScore – AwayTeamScore)
36	O2	WinLoss**	Whether the home team wins or loses the game

* Over/Under: Whether a team will go over or under the expected score difference

** Output variables: ScoreDiff for regression models and WinLoss for binary classification models

I1: Offense/defense; I2: Game outcome; I3: Team configuration; I4: Against the odds; I5: Conference stats

ID: Identifier variables; O1: Output variable for regression models; O2: Output variable for classification models

Methods

In this study, three popular prediction techniques were used to build models (and to compare them to each other): artificial neural networks, decision trees, and support vector machines. These prediction techniques were selected based on their ability to model both classification- and regression-type prediction problems and their popularity in recently published data mining literature. More details about these popular data mining methods can be found in Chapter 5 in this book.

Evaluation

In order to compare predictive accuracy of all models to one another, the researchers used a stratified *k*-fold cross-validation methodology. In a stratified version of *k*-fold cross-validation, the folds are created such that they contain approximately the same proportion of predictor labels (i.e., classes) as the original data set. In this study, the value of *k* was set to 10 (i.e., the complete set of 244 samples was split into 10 subsets, each having about 25 samples), which is a common practice in predictive data mining applications. A graphical depiction of the 10-fold cross-validation was shown earlier in this chapter (see Figure 4.11). To compare the prediction models that were developed using the previously mentioned three data mining techniques, the researchers chose to use three common performance criteria: accuracy, sensitivity, and specificity. The simple formulas for these metrics were also explained earlier in this chapter.

Results

The prediction results of the three modeling techniques are presented in Table 4.5 and Table 4.6. Table 4.5 presents the 10-fold cross-validation results of the classification methodology where the three data mining techniques are formulated to have a binary-nominal output variable (i.e., WinLoss). Table 4.6 presents the 10-fold cross-validation results of the regression-based classification methodology where the three data mining techniques are formulated to have a numerical output variable (i.e., ScoreDiff). In the regression-based classification prediction, the numerical output of the models is converted to a classification type by labeling the positive WinLoss numbers with *Win* and negative WinLoss numbers with *Loss* and then tabulating them in the confusion matrices. Using the confusion matrices, the overall prediction accuracy, sensitivity, and specificity of each model type are calculated and presented in these two tables.

As the results indicate, the classification-type prediction methods performed better than regression-based classification-type prediction methodologies. Among the three data mining technologies, classification and regression trees produced better prediction accuracy in both prediction methodologies. Overall, classification and regression tree classification models produced a 10-fold cross-validation accuracy of 86.48%, followed by support vector machines (with a 10-fold cross-validation accuracy of 79.51%) and neural networks (with a 10-fold cross-validation accuracy of 75.00%). Using a *t*-test, the researchers found that these accuracy values were significantly different at the 0.05 alpha level; that is, the decision tree was a significantly better predictor of this domain than were the neural network and support vector machine; and the support vector machine was a significantly better predictor than the neural networks.

Table 4.5 Prediction Results for the Direct Classification Methodology

Prediction Method (Classification*)		Confusion Matrix Win	Loss	Accuracy** (in %)	Sensitivity (in %)	Specificity (in %)
ANN (MLP)	Win	92	42	75.00	68.66	82.73
	Loss	19	91			
SVM (RBF)	Win	105	29	79.51	78.36	80.91
	Loss	21	89			
DT (C&RT)	Win	113	21	**86.48**	84.33	89.09
	Loss	12	98			

* The output variable is a binary categorical variable (Win or Loss).

** Differences were significant ($p < 0.01$).

Table 4.6 Prediction Results for the Regression-Based Classification Methodology

Prediction Method (Regression-Based*)		Confusion Matrix Win	Loss	Accuracy** (in %)	Sensitivity (in %)	Specificity (in %)
ANN (MLP)	Win	94	40	72.54	70.15	75.45
	Loss	27	83			
SVM (RBF)	Win	100	34	74.59	74.63	74.55
	Loss	28	82			
DT (C&RT)	Win	106	28	77.87	76.36	79.10
	Loss	26	84			

* The output variable is a numerical/integer variable (ScoreDiff).

** Differences were significant ($p < 0.01$).

The results of the study showed that the classification-type models predicted the game outcomes better than did regression-based classification models. Even though these results are specific to the application domain and the data used in this study, and therefore should not be generalized beyond the scope of the study, they are exciting because decision trees are not only the best predictors but also the best for understanding and deployment, compared to the other two machine learning techniques employed in this study. More details about this study can be found in Delen et al. (2012).

Source: Delen, D., D. Cogdell, & N. Kasap, "A Comparative Analysis of Data Mining Methods in Predicting NCAA Bowl Outcomes," *International Journal of Forecasting*, 28 (2012): 543–552.

References

Bhandari, I., et al. (1997). "Advanced Scout: Data Mining and Knowledge Discovery in NBA Data," *Data Mining and Knowledge Discovery*, 1(1): 121–125.

Buck, N. (2000/2001, December/January). "Eureka! Knowledge Discovery," *Software Magazine*.

Chan, P. K., W. Phan, A. Prodromidis, & S. Stolfo. (1999). "Distributed Data Mining in Credit Card Fraud Detection," *IEEE Intelligent Systems*, 14(6): 67–74.

CRISP-DM. (2013). *Cross-Industry Standard Process for Data Mining (CRISP-DM)*. www.the-modeling-agency.com/crisp-dm.pdf (accessed February 2013).

Davenport, T. H. (2006, January). "Competing on Analytics," *Harvard Business Review*.

Delen, D., D. Cogdell, & N. Kasap. (2012). "A Comparative Analysis of Data Mining Methods in Predicting NCAA Bowl Outcomes," *International Journal of Forecasting*, 28: 543–552.

Delen, D., R. Sharda, & P. Kumar. (2007). "Movie Forecast Guru: A Web-Based DSS for Hollywood Managers," *Decision Support Systems*, 43(4): 1151–1170.

Nemati, H. R., & C. D. Barko. (2001). "Issues in Organizational Data Mining: A Survey of Current Practices," *Journal of Data Warehousing*, 6(1): 25–36.

Quinlan, J. R. (1986). "Induction of Decision Trees," *Machine Learning*, 1: 81–106.

Sharda, R., & D. Delen. (2006). "Predicting Box-Office Success of Motion Pictures with Neural Networks," *Expert Systems with Applications*, 30: 243–254.

Stekler, H. O., D. Sendor, & R. Verlander. (2010). "Issues in Sports Forecasting," *International Journal of Forecasting*, 26: 606–621.

Van Bruggen, G. H., M. Spann, G. L. Lilien, & B. Skiera. (2010). "Prediction Markets as Institutional Forecasting Support Systems," *Decision Support Systems*, 49: 404–416.

Wald, M. L. (2004, February 21). "U.S. Calls Release of JetBlue Data Improper," *The New York Times*.

Wilson, R., & R. Sharda. (1994). "Bankruptcy Prediction Using Neural Networks," *Decision Support Systems*, 11: 545–557.

Wright, C. (2012). *Statistical Predictors of March Madness: An Examination of the NCAA Men's' Basketball Championship*. http://economics-files.pomona.edu/GarySmith/Econ190/Wright%20March%20Madness%20Final%20Paper.pdf (accessed February 2013).

Zaima, A. (2003). "The Five Myths of Data Mining," *What Works: Best Practices in Business Intelligence and Data Warehousing*, 15: 42–43.

5

Data Mining Algorithms

Data mining methods, especially the ones used to build predictive models, are derived from either traditional statistical techniques/algorithms or from contemporary machine learning techniques/systems. While statistical techniques are well established and theoretically sound, machine learning techniques are more practical and accurate (according to a wealth of comparative studies published in the literature). The statistical techniques that made the biggest impact in the evolution of data mining include discriminant analysis, linear regression, and logistic regression. The most popular machine learning techniques used in numerous successful data mining projects include nearest neighbor, artificial neural networks, and support vector machines. All three of these machine learning techniques can handle both classification- as well as regression-type prediction problems. Often, they are applied to complex prediction problems where other techniques are not capable of producing satisfactory results.

Because of the popularity of machine learning techniques in the data mining literature, in this chapter we explain these techniques in detail but without getting too technical in their algorithmic background. We also briefly explain the previously mentioned statistical techniques to provide balanced coverage of the technical spectrum of data mining.

Nearest Neighbor

Data mining algorithms tend to be highly mathematical and computationally intensive. Two popular ones—artificial neural networks and support vector machines—are involved in time-demanding, computationally intensive iterative mathematical derivations. In contrast, the *k*-nearest neighbor algorithm (*k*-NN) seems overly simplistic for a competitive prediction method. It is easy to understand (and explain to others) what it does and how it does it. *k*-NN is a prediction method for classification as well as regression-type prediction problems. *k*-NN is a type of instance-based learning (or lazy learning) where the function is only approximated locally, and all computations are deferred until the actual prediction.

The *k*-NN algorithm is among the simplest of all machine learning algorithms. For instance, in classification-type prediction, a case is classified by a majority vote of its neighbors, with the object being assigned to the class most common among its *k* nearest neighbors (where *k* is a positive integer). If $k = 1$, then the case is simply assigned to the class of its nearest neighbor. To illustrate the concept with an example, Figure 5.1 shows a simple two-dimensional space representing the values for the two variables x and y; the star represents a new case (or object), and circles and squares represent known cases (or examples). The task is to assign the new case either to circles or squares, based on its closeness (similarity) to one or the other. If you set the value of *k* to 1 ($k = 1$), the assignment should be made to square because the closest example to a star is a square. If you set the value of *k* to 3 ($k = 3$), then the assignment should be made to a circle because there are two circles and one square, and hence by simple majority vote rule, the circle gets the assignment of the new case. Similarly, if you set the value of *k* to 5 ($k = 5$), then the assignment should be made to the square class. This overly simplified example is meant to illustrate the importance of the value assigned to *k*.

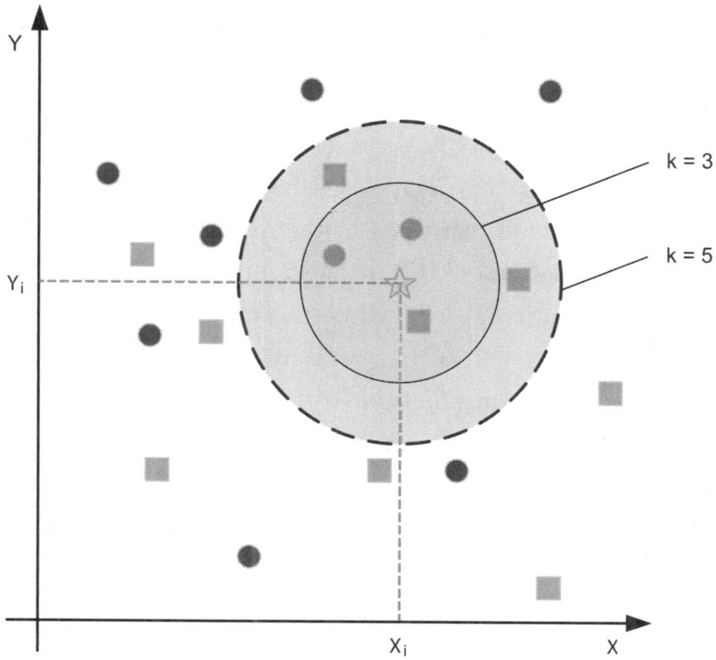

Figure 5.1 The Importance of the Value of *k* in the *k*-NN Algorithm

The same method can also be used for regression-type prediction tasks, by simply averaging the values of the k nearest neighbors and assigning that result to the case being predicted. It can be useful to weight the contributions of the neighbors so that the nearer neighbors contribute more to the average than do the more distant ones. A common weighting scheme is to give each neighbor a weight of $1/d$, where d is the distance to the neighbor. This scheme is essentially a generalization of linear interpolation.

The neighbors are taken from a set of cases for which the correct classification (or, in the case of regression, the numeric value of the output value) is known. This can be thought of as the training set for the algorithm, even though no explicit training step is required. The k-nearest neighbor algorithm is sensitive to the local structure of the data.

Similarity Measure: The Distance Metric

One of the two critical decisions that an analyst has to make while using k-NN is to determine the similarity measure; the other is to determine the value of k, which is explained below. In the k-NN algorithm, the similarity measure is a mathematically calculable distance metric. Given a new case, k-NN makes predictions based on the outcome of the k neighbors closest in distance to that point. Therefore, to make predictions with k-NN, an analyst needs to define a metric for measuring the distance between the new case and the cases from the examples. One of the most popular choices for measuring this distance is known as Euclidean (Equation 3), which is simply the linear distance between two points in a dimensional space; the other popular one is the rectilinear (a.k.a. city-block or Manhattan distance) (Equation 2). Both of these distance measures are special cases of Minkowski distance (Equation 1).

Here is the equation for finding the Minkowski distance:

$$d(i,j)=\sqrt[q]{(|x_{i1}-x_{j1}|^q+|x_{i2}-x_{j2}|^q+\ldots+|x_{ip}-x_{jp}|^q)} \quad (1)$$

where $i = (x_{i1}, x_{i2}, \ldots, x_{ip})$ and $j = (x_{j1}, x_{j2}, \ldots, x_{jp})$ are two p-dimensional data objects (e.g., a new case and an example in the data set), and q is a positive integer.

If $q = 1$, then d is called *Manhattan distance* and is found using this equation:

$$d(i,j)=|x_{i1}-x_{j1}|+|x_{i2}-x_{j2}|+\ldots+|x_{ip}-x_{jp}| \quad (2)$$

If $q = 2$, then d is called Euclidean distance and is found using this equation:

$$d(i,j)=\sqrt{(|x_{i1}-x_{j1}|^2+|x_{i2}-x_{j2}|^2+\ldots+|x_{ip}-x_{jp}|^2)} \quad (3)$$

Obviously, these measures apply only to numerically represented data. What happens with nominal data? There are ways to measure distance for nonnumeric data as well. In the simplest case, for a multivalue nominal variable, if the value of that variable for the new case and the example case is the same, the distance would be zero; otherwise, it would be one. In cases such as text classification, more sophisticated metrics exist, such as the overlap metric (or Hamming distance). Often, the classification accuracy of k-NN can be improved significantly if the distance metric is determined through an experimental design where different metrics are tried and tested to identify the best one for the given problem.

Parameter Selection

The best choice for k depends on the data; generally, larger values of k reduce the effect of noise on the classification (or regression) but also make boundaries between classes less distinct. An "optimal" value of k can be found by using some heuristic techniques, such as cross-validation. The special case where the class is predicted to be the class of the closest training sample (i.e., when $k = 1$) is called the *nearest-neighbor algorithm*.

Cross-Validation

Cross-validation is a well-established experimentation technique that can be used to determine optimal values for a set of unknown model parameters. It applies to most, if not all, of the machine learning techniques, where there are a number of model parameters to be determined. The general idea of this experimentation method is to divide the data sample into a number of randomly drawn, disjointed subsamples (i.e., v number of folds). For each potential value of k, the k-NN model is used to make predictions on the vth fold while using the $v - 1$ folds as the examples, and evaluate the error. The common choice for this error is the root-mean-squared-error (RMSE) for regression-type predictions and percentage of correctly classified instances (i.e., hit rate) for classification-type predictions. This process of testing each fold against the remaining of examples repeats v times. At the end of the v number of cycles, the computed errors are accumulated to yield a goodness measure of the model (i.e., how well the model predicts with the current value of the k). At the end, the k value that produces the smallest overall error is chosen as the optimal value for that problem. Figure 5.2 shows a simple process where training data is used to determine optimal values for k and distance metric, which are then used to predict new incoming cases.

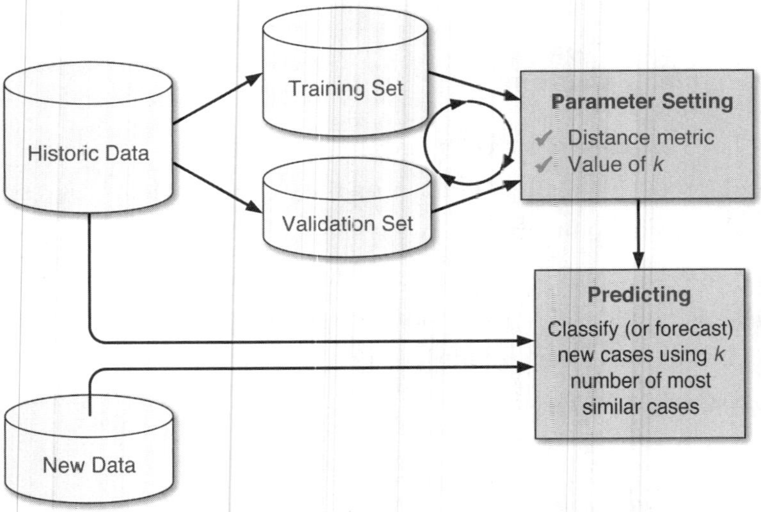

Figure 5.2 The Process of Determining the Optimal Values for the Distance Metric and *k*

As in the simple example given above, the accuracy of the k-NN algorithm can be significantly different with different values of k. Furthermore, the predictive power of the k-NN algorithm degrades in the presence of noisy, inaccurate, or irrelevant features. Much research effort has been put into feature selection and normalization/scaling to ensure reliable prediction results. A particularly popular approach is the use of evolutionary algorithms (e.g., genetic algorithms) to optimize the set of features included in the k-NN prediction system. In binary (two-class) classification problems, it is helpful to choose an odd number for k to avoid tied votes.

A drawback to the basic majority voting classification in k-NN is that the classes with the more frequent examples tend to dominate the prediction of the new vector, as they tend to come up in the k nearest neighbors when the neighbors are computed due to their large number. One way to overcome this problem is to weigh the classification, taking into account the distance from the test point to each of its k nearest neighbors. Another way to overcome this drawback is to use one level of abstraction in data representation.

The naïve version of the algorithm is easy to implement by computing the distances from the test sample to all stored vectors, but it is computationally intensive, especially when the size of the training set grows. Many nearest-neighbor search algorithms have been proposed over the years; they generally seek to reduce the number of distance evaluations actually performed. Using an appropriate nearest-neighbor search algorithm makes k-NN computationally tractable even for large data sets.

Artificial Neural Networks

Neural networks use a brain metaphor for information processing. These models are biologically inspired rather than an exact replica of how the brain actually functions. Neural networks have been shown to be very promising systems in many forecasting and business classification applications due to their ability to "learn" from the data, their nonparametric nature (i.e., no rigid assumptions), and their ability to generalize. *Neural computing* refers to a pattern-recognition methodology for machine learning. The resulting model from neural computing is often called an *artificial neural network* (*ANN*) or a *neural network*. Neural networks have been used in many business applications for pattern recognition, forecasting, prediction, and classification. Neural network computing is a key component of any data mining toolkit. Applications of neural networks abound in finance, marketing, manufacturing, operations, information systems, and so on. Therefore, we devote this chapter to developing a better understanding of neural network models, methods, and applications.

The human brain possesses astonishing capabilities for information processing and problem solving that even the most advanced computers of this era cannot compete with in many aspects. It has been claimed that a model or a system that is enlightened and supported by the results from brain research, with a structure similar to that of biological neural networks, could exhibit similar intelligent functionality. Based on this bottom-up approach, ANNs (also known as connectionist models, parallel distributed processing models, neuromorphic systems, or simply neural networks) have been developed as biologically inspired and plausible models for various tasks.

Biological neural networks are composed of many massively interconnected neurons. Each neuron possesses axons and dendrites, fingerlike projections that enable the neuron to communicate with its neighboring neurons by transmitting and receiving electrical and chemical signals. More-or-less resembling the structure of their biological counterparts, ANNs are composed of interconnected, simple processing elements called artificial neurons. When processing information, the processing elements in an ANN operate concurrently and collectively, much as in biological neurons. ANNs possess some desirable traits similar to those of biological neural networks, such as the abilities to learn, self-organize, and support fault tolerance.

For more than half a century, researchers have been investigating ANNs. The formal study of ANNs began with the pioneering work of McCulloch and Pitts in 1943. Inspired by the results of biological experiments and observations, McCulloch and Pitts (1943) introduced a simple model of a binary artificial neuron that captured some of the functions of biological neurons. Using information-processing machines to model the brain, McCulloch and Pitts built their neural network model using a large number of interconnected artificial binary neurons. From these beginnings, neural network research became quite popular in the late 1950s and early 1960s. After a thorough analysis of an early neural network model (called the perceptron, which used no hidden layer) as well as a pessimistic evaluation of the research potential by Minsky and Papert in 1969, interest in neural networks diminished.

During the past two decades, there has been an exciting resurgence in ANN studies due to the introduction of new network topologies, new activation functions, and new learning algorithms, as well as progress in neuroscience and cognitive science. Advances in theory and methodology have overcome many of the obstacles that hindered neural network research a few decades ago. Evidenced by the appealing results of numerous studies, neural networks are gaining acceptance and popularity. In addition, the desirable features in neural information processing make neural networks attractive for solving complex problems. ANNs have been applied to numerous complex problems in a variety of application settings. The successful use of neural network applications has inspired renewed interest from industry and business.

Biological and Artificial Neural Networks

The human brain is composed of special cells called *neurons*. These cells do not die and replenish when a person is injured. (All other cells reproduce to replace themselves and then die.) This phenomenon may explain why humans retain information for an extended period of time and start to lose it when they get old—as the brain cells gradually start to die. Information storage spans sets of neurons. The brain has anywhere from 50 billion to 150 billion neurons, of which there are more than 100 different kinds. Neurons are partitioned into groups called *networks*. Each network contains several thousand highly interconnected neurons. Thus, the brain can be viewed as a collection of neural networks.

The ability to learn and to react to changes in our environment requires intelligence. The brain and the central nervous system control thinking and intelligent behavior. People who suffer brain damage have difficulty learning and reacting to changing environments. Even so, undamaged parts of the brain can often compensate with new learning.

A portion of a network composed of two cells is shown in Figure 5.3. The cell itself includes a nucleus (the central processing portion of the neuron). To the left of the first cell, the dendrites provide input signals to the cell. To the right, the axon sends output signals to the second cell via the axon terminals. These axon terminals merge with the dendrites of the second cell. Signals can be transmitted unchanged, or they can be altered by synapses. A synapse is able to increase or decrease the strength of the connection between neurons and cause excitation or inhibition of a subsequent neuron. This is how information is stored in the neural networks.

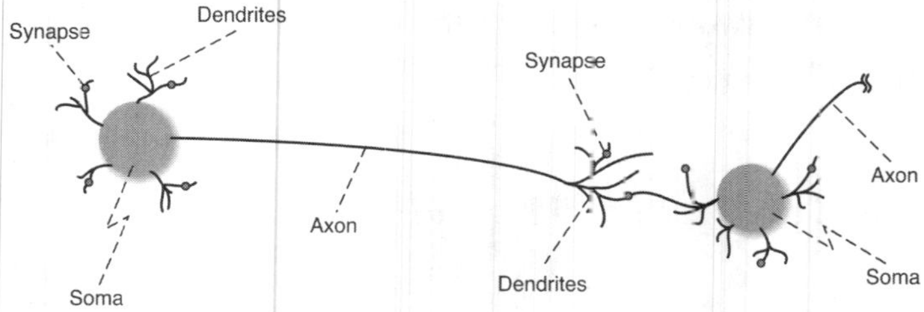

Figure 5.3 A Biological Neural Network: Two Interconnected Cells/Neurons

An ANN emulates a biological neural network. Neural computing actually uses a very limited set of concepts from biological neural systems. A simple terminology mapping between biological and artificial neural networks is shown in Table 5.1. It is more of an analogy to the human brain than an accurate model of it. Neural concepts usually are implemented as software simulations of the massively parallel processes involved in processing interconnected elements (also called artificial neurons, or neurodes) in a network architecture. The artificial neuron receives inputs analogous to the electrochemical impulses that dendrites of biological neurons receive from other neurons. The output of the artificial neuron corresponds to signals sent from a biological neuron over its axon. These artificial signals can be changed by weights in a manner similar to the physical changes that occur in the synapses (see Figure 5.4)

Table 5.1 Mapping Between Biological and Artificial NNs

Biological NN	Artificial NN
Soma	Node
Dendrites	Input
Axon	Output
Synapse	Weight
Slow	Fast
Many neurons (109)	Few neurons (a dozen to hundreds of thousands)

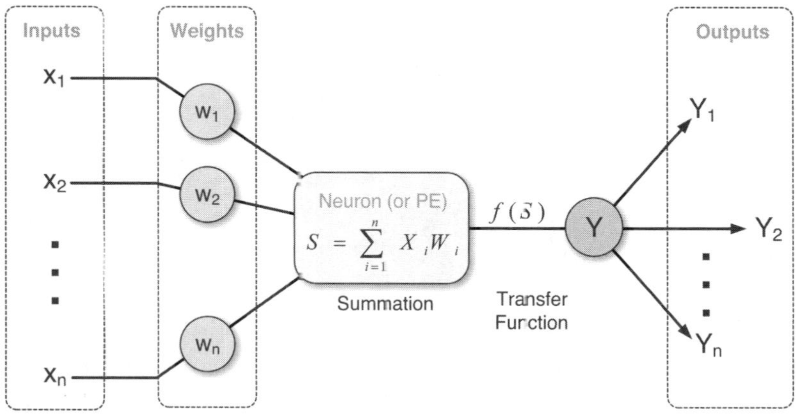

Figure 5.4 Processing Information in an Artificial Neuron

Several ANN paradigms have been proposed for applications in a variety of problem domains. Perhaps the easiest way to differentiate among the various neural models is on the basis of how they structurally emulate the human brain, the way they process information, and how they learn to perform their designated tasks.

Because they are biologically inspired, the main processing elements of a neural network are individual neurons, analogous to the brain's neurons. These artificial neurons receive information from other neurons or external input stimuli, perform a transformation on the inputs, and then pass on the transformed information to other neurons or external outputs. This is similar to how it is presently thought that the human brain works. Passing information from neuron to neuron can be thought of as a way to activate, or trigger, a response from certain neurons based on the information or stimulus received.

How information is processed by a neural network is inherently a function of its structure. Neural networks can have one or more layers of neurons. These neurons can be highly or fully interconnected, or only certain layers can be connected. Connections between neurons have an associated weight. In essence, the "knowledge" possessed by the network is encapsulated in these interconnection weights. Each neuron calculates a weighted sum of the incoming neuron values, transforms this input, and passes on its neural value as the input to subsequent neurons. Typically, although not always, this input/output transformation process at the individual neuron level is performed in a nonlinear fashion.

Basic Components of an ANN

A neural network is composed of processing elements that are organized in different ways to form the network's structure. The basic processing unit is the neuron. A number of neurons are organized into a network. Neurons can be organized in a number of different ways; these various network patterns are referred to as *topologies*. One popular approach, known as the feed-forward/back-propagation paradigm (or simply back-propagation), allows all neurons to link the output in one layer to the input of the next layer, but it does not allow any feedback linkage (Haykin, 2009). Back-propagation is the most commonly used network paradigm.

Processing Elements

The processing elements (PEs) of an ANN are artificial neurons. Each neuron receives inputs, processes them, and delivers a single output, as shown in Figure 5.4. The input can be raw input data or the output of other processing elements. The output can be the final result (e.g., 1 means yes, 0 means no), or it can be input to other neurons.

Network Structure

Each ANN is composed of a collection of neurons that are grouped into layers. A typical structure is shown in Figure 5.5. Note the three layers: input, intermediate (called the hidden layer), and output. A hidden layer is a layer of neurons that takes input from the previous layer and converts those inputs into outputs for further processing. Several hidden layers can be placed between the input and output layers, although it is common to use only one hidden layer. In that case, the hidden layer simply converts inputs into a nonlinear combination and passes the transformed inputs to the output layer. The most common interpretation of the hidden layer is as a feature-extraction mechanism; that is, the hidden layer converts the original inputs in the problem into a higher-level combination of such inputs.

There are several neural network structures (or architectures) for a variety of analytics problems. The most common ones include feed-forward (multilayer perceptron with back-propagation), associative memory, recurrent networks, Kohonen's self-organizing feature maps, and Hopfield networks. Perhaps the most commonly used one is the feed-forward neural network with a multilayered perceptron (commonly known as MLP). The generic architecture of a feed-forward network is shown in Figure 5.5, where the information flows unidirectionally from the input layer to hidden layers and from hidden layers to the output layer.

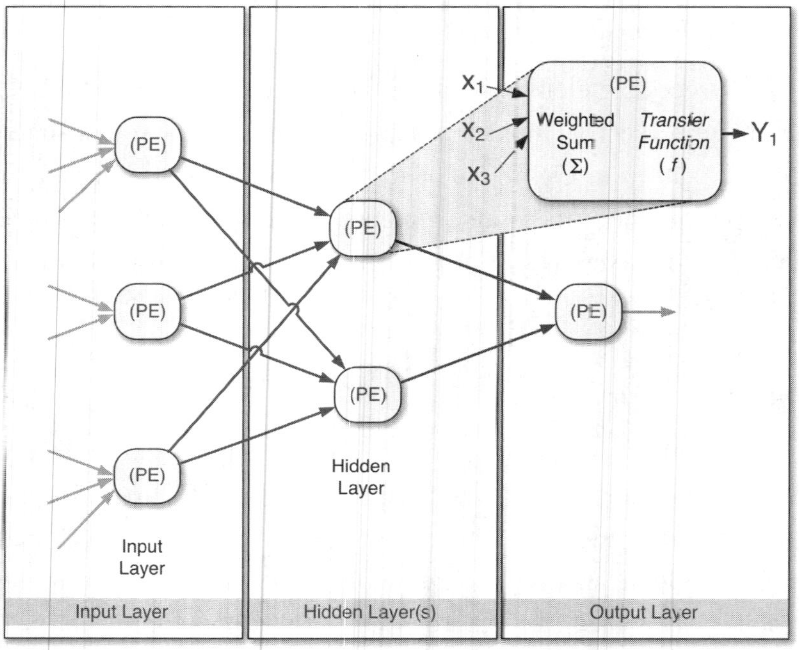

Figure 5.5 Structure of a Feed-Forward Neural Network

Like a biological network, an ANN can be organized in several different ways (i.e., topologies or architectures); that is, the neurons can be interconnected in different ways. When information is processed, many of the processing elements perform their computations at the same time. This parallel processing resembles the way the brain works, and it differs from the serial processing of conventional computing.

How Does the Learning Happen?

MLP architecture in ANN uses supervised learning, where the learning process is inductive; that is, the values of parameters (i.e., the connection weights) are derived from existing cases. The usual process of learning in MLP-type ANNs involves three tasks (see Figure 5.6):

1. Compute temporary outputs.
2. Compare outputs with desired targets.
3. Adjust the weights and repeat the process.

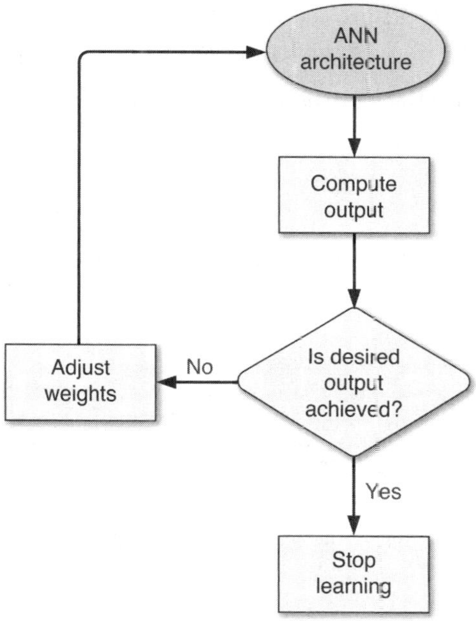

Figure 5.6 The Supervised Learning Process of an ANN

Back-Propagation

Back-propagation (or back-error propagation) is the most widely used supervised learning algorithm in neural networks. It is relatively easy to implement and produces surprisingly accurate results. A back-propagation network includes one or more hidden layers. This type of network is considered feed-forward because there are no interconnections between the output of a processing element and the input of a node in the same layer or in a preceding layer. Externally provided correct patterns are compared with the neural network's output during (supervised) training, and feedback is used to adjust the weights until the network has categorized all the training patterns as correctly as possible. (The error tolerance is set in advance.)

Starting with the output layer, errors between the actual and desired outputs are used to correct the weights for the connections to the previous layer (see Figure 5.7). For any output neuron, the error is calculated as the difference between the network output (Y) and the actual outputs (Z). Using some type of nonlinear function, this error is propagated back to the network weight for proper adjustment (increase or decrease).

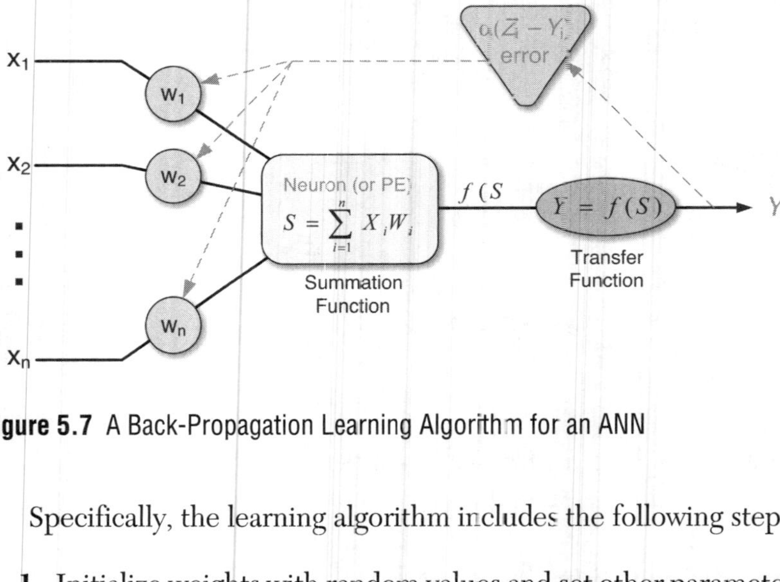

Figure 5.7 A Back-Propagation Learning Algorithm for an ANN

Specifically, the learning algorithm includes the following steps:

1. Initialize weights with random values and set other parameters.
2. Read in the input vector and the desired output.
3. Compute the actual output via the calculations, working forward through the layers.
4. Compute the error.
5. Change the weights by working backward from the output layer through the hidden layers.

This procedure is repeated for the entire set of input vectors until the desired output and the actual output agree to within some predetermined tolerance. Given the calculation requirements for one iteration, a large network can take a very long time to train; therefore, in one variation, a set of cases is run forward, and an aggregated error is fed backward to speed up learning. Sometimes, depending on the initial random weights and network parameters, the network does not converge to a satisfactory performance level. When this is the case, new random weights must be generated, and the network parameters, or even its structure, may have to be modified before another attempt is made.

Sensitivity Analysis in Artificial Neural Networks

Neural networks have been used as effective tools for solving highly complex real-world problems in a wide range of application areas. Even though ANNs have been proven in many problem scenarios to be superior predictors and/or cluster identifiers (compared to their traditional counterparts), in some applications, there exists an additional need to know how the ANN does what it does. ANNs are typically thought of as black boxes, capable of solving complex problems but lacking explanations for their capabilities. This phenomenon is commonly referred to as *black-box syndrome*.

It is important to be able to explain a model's "inner being"; such an explanation offers assurance that the network has been properly trained and will behave as desired once deployed in a business intelligence environment. The need to "look under the hood" might be attributable to a relatively small training set (as a result of the high cost of data acquisition) or a very high liability in the case of a system error. One example of such an application is the deployment of airbags in automobiles. Here, both the cost of data acquisition (crashing cars) and the liability concerns (danger to human lives) are rather significant. Another representative example for the importance of explanation is loan-application processing. If an applicant is refused for a loan, the applicant has a right to know why. Having a prediction system that does a good job differentiating good applications from bad ones may not be sufficient if it does not also provide justification for its decisions.

A variety of techniques have been proposed for analysis and evaluation of trained neural networks. These techniques provide a clear interpretation of how a neural network does what it does—that is, specifically how (and to what extent) the individual inputs factor into the generation of specific network output. Sensitivity analysis has been the front-runner among the techniques proposed for shedding light into the black-box characterization of trained neural networks.

Sensitivity analysis is a method for extracting the cause-and-effect relationships among the inputs and outputs of a trained neural network model. In the process of performing sensitivity analysis, the trained neural network's learning capability is disabled so that the network weights are not affected. The basic procedure involved in sensitivity analysis is to systematically perturb the inputs to the network within the allowable value ranges and to record the corresponding changes in the output for each and every input variable. Figure 5.8 shows a graphical illustration of this process. The first input is varied between its mean plus and minus a user-defined number of standard deviations (or, for categorical variables, all the possible values are used), while all other input variables are fixed at their respective means (or modes). The network output is computed for a user-defined number of steps above and below the mean. This process is repeated for each input. As a result, a report is generated that summarizes the variation of each output with respect to the variation in each input. The generated report often contains a column plot (where numeric importance values are presented on the y-axis and input variables are listed in the x-axis) reporting the relative sensitivity/importance values for each input variable.

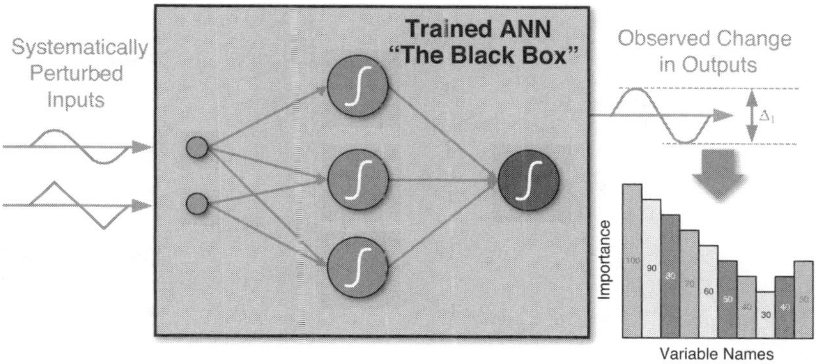

Figure 5.8 An Illustration of Sensitivity Analysis on an ANN Model

Support Vector Machines

Support vector machines (SVMs) are among the most popular machine learning techniques in the recent past, largely because of their superior predictive performance and their theoretical foundation. SVMs are among the supervised learning methods that produce input/output functions from a set of labeled training data. The function between the input and output vectors can be either a classification function (used to assign cases into predefined classes) or a regression function (used to estimate the continuous numeric value of the desired output). For classification, nonlinear kernel functions are often used to transform the input data (naturally representing highly complex nonlinear relationships) to a high dimensional feature space in which the input data becomes linearly separable. Then, the maximum-margin hyperplanes are constructed to optimally separate the output classes from each other in the training data.

For a given classification-type prediction problem, generally speaking, many linear classifiers (hyperplanes) can separate the data into multiple subsections, each representing one of the classes (see Figure 5.9a, where the two classes are represented with circles [●] and squares [■]). However, only one hyperplane achieves the maximum separation between the classes (see Figure 5.9b, where the hyperplane and the two maximum-margin hyperplanes are separating the two classes).

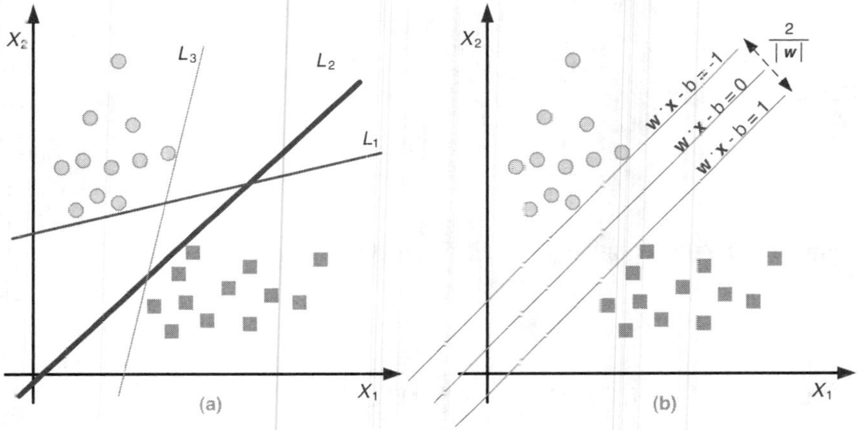

Figure 5.9 Separation of the Two Classes with Support Vector Machines

Data used in SVMs may have more than two dimensions (i.e., two distinct classes). In that case, we would be interested in separating data using $n - 1$ dimensional hyperplanes, where n is the number of dimensions (i.e., class labels). This may be seen as a typical form of linear classifier, where we are interested in finding $n - 1$ hyperplanes so that the distance from the hyperplanes to the nearest data points are maximized. The assumption is that the larger the margin or distance between these parallel hyperplanes, the better the generalization power of the classifier (i.e., prediction power of the SVM model). If such hyperplanes exist, they can be mathematically represented using quadratic optimization modeling. These hyperplanes are known as the maximum-margin hyperplanes, and a linear classifier is known as a maximum-margin classifier.

In addition to their solid mathematical foundation in statistical learning theory, SVMs have also demonstrated highly competitive performance in numerous real-world prediction problems, such as medical diagnosis, bioinformatics, face/voice recognition, demand forecasting, image processing, and text mining, which has established SVMs as one of the most popular analytics tools for knowledge discovery and data mining. Similarly to artificial neural networks, SVMs can be used to approximate any multivariate function to any desired degree of accuracy. Therefore, they are of particular interest in modeling highly nonlinear, complex problems, systems, and processes.

The Process of Building SVM Models

Recently, using support vector machines has become a popular technique for classification-type problems because of their predictive accuracy and self-expandability. Even though people consider them easier to use than artificial neural networks, users who are not familiar with the intricacies of SVMs often get unsatisfactory results. In this section we provide a simple process-based approach to model building with and use of SVMs that is likely to produce good results. A pictorial representation of the three-step process is given in Figure 5.10.

Following are short descriptions of the steps in this process.

Step 1: Preprocess the Data

Because real-world data is anything but perfect, a data mining analyst needs to do the due diligence to scrub and transform the data for SVM. As is the case with any other data mining methods and algorithms, this step is likely to be the most time-demanding and least enjoyable, yet most essential, part of working with SVMs. Some of the tasks in this step include handling missing, incomplete, and noisy values and numerisizing, normalizing, and standardizing the data:

- **Numerisizing the data.** SVMs require that each data instance be represented as a vector of real numbers. Hence, if there are categorical attributes, an analyst has to first convert them into numeric data. A common recommendation is to use m pseudo binary variables to represent an m-class attribute (where $m \geq 3$).

In practice, only one of the m variables assumes the value of 1, and others assume the value of 0, based on the actual class of the case; this is also called 1-of-m representation. For example, a three-category attribute such as {red, green, blue} can be represented as (0,0,1), (0,1,0), and (1,0,0).

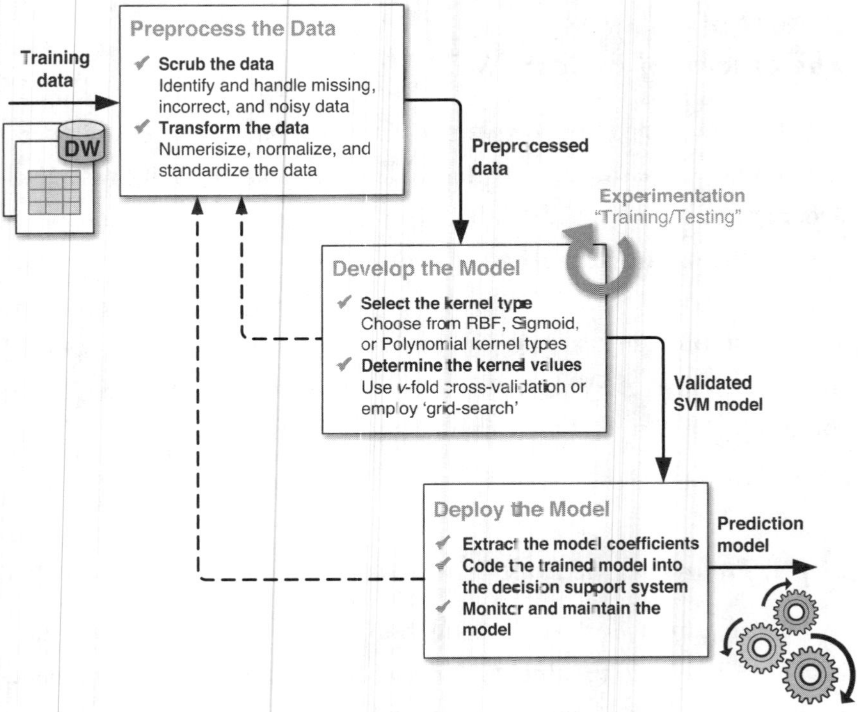

Figure 5.10 A Simple Process for Developing and Using SVM Models

- **Normalizing the data.** As is the case with artificial neural networks, SVMs also require normalization and/or scaling of numeric values. The main advantage of normalization is to avoid having attributes in greater numeric ranges dominate those in smaller numeric ranges. Another advantage is that it helps perform numeric calculations during the iterative process of model building. Because kernel values usually depend on the inner products of feature vectors (e.g., the linear kernel, the polynomial kernel), large attribute values might slow the training process. The usual recommendation is to normalize each attribute to the range [−1, +1] or [0, 1]. Of course, an analyst has to use the same normalization method to scale testing data before testing.

Step 2: Develop the Model

Once the data is preprocessed, the model building step starts. Compared to the other two steps, this is the most enjoyable part of the process, where the prediction model comes alive. Because SVM has a number of parameter options, the process requires a lengthy process of determining the optimal combination of those parameters for the best possible performance. The most important of those parameters are the kernel type and consequently kernel-related subparameters.

There are four common kernels, and an analyst must decide which one to use (or whether to try them all, one at a time, using a simple experimental design approach). Once the kernel type is selected, the analyst needs to select the value of penalty parameter C and the kernel parameters. Generally speaking, RBF is a reasonable first choice for the kernel type. The RBF kernel aims to nonlinearly map data into a higher dimensional space, and this (unlike with a linear kernel) handles the cases where the relationship between input and output vectors is highly nonlinear. Besides, the linear kernel is just a special case of the RBF kernel. There are two parameters to choose for RBF kernels: C and γ. It is not known beforehand which C and γ are the best for a given prediction problem; therefore, some kind of parameter search method needs to be used. The goal for the search is to identify optimal values for C and γ so that the classifier can accurately predict unknown data (i.e., testing data). The two most commonly used search methods are cross-validation and grid search.

Step 3: Deploy the Model

Once an optimal SVM prediction model is developed, the next step is to integrate it into the decision support system. For that, there are two options: (1) converting the model into a computational object (e.g., a Web service, a Java Bean, or a COM object) that takes the input parameter values and provides output prediction or (2) extracting the model coefficients and integrating them directly into the decision support system. The SVM models are useful (i.e., accurate and actionable) only if the behavior of the underlying domain stays the same. For some reason, if the behavior changes, so does the accuracy of the model. Therefore, it's important to continuously assess the performance of the models and decide when they are no longer accurate and, hence, need to be retrained.

Support Vector Machines Versus Artificial Neural Networks

Even though some people characterize SVMs as a special case of ANNs, most recognize these as two competing machine learning techniques with different qualities. Let's look at a few points that help SVMs stand out against ANNs. Historically, the development of ANNs followed a heuristic path, with applications and extensive experimentation preceding theory. In contrast, the development of SVMs involved sound statistical learning theory first, followed by implementation and experiments. A significant advantage of SVMs is that while ANNs may suffer from multiple local minima, the solutions to SVMs are global and unique. Two more advantages of SVMs are that they have simple geometric interpretations and give sparse solutions. The reason that SVMs often outperform ANNs in practice is that they successfully deal with the "overfitting" problem, which is a big issue with ANNs.

Besides the above-mentioned advantages of SVMs (from a practical point of view), they also have some limitations. An important issue that is not entirely solved is the selection of the kernel type and kernel function parameters. Other, perhaps more important, limitations of SVMs are the speed and size, both in training and testing cycles. Model building with SVMs involves complex and time-demanding calculations. From a practical point of view, perhaps the most serious problems with SVMs are the high algorithmic complexity and extensive memory requirements of the required quadratic programming in large-scale tasks. Despite these limitations, because SVMs are based on a sound theoretical foundation and the solutions they produce are global and unique in nature (as opposed to getting stuck in suboptimal alternatives as local minima do), today's SVMs are some of the most popular prediction modeling techniques in the data mining arena. Their use and popularity will only increase as the popular commercial data mining tools start to incorporate them into their modeling arsenals.

Linear Regression

Regression, especially linear regression, is perhaps the most widely used analysis technique in statistics. Historically speaking, the roots of regression date back to the 1920s and 1930s, to the early work on inherited characteristics of sweet peas by Sir Francis Galton and subsequently Karl Pearson. Since then, regression has become the statistical technique for characterization of relationships between explanatory (input) variable(s) and response (output) variable(s).

As popular as it is, regression is essentially a relatively simple statistical technique to model the dependence of a variable (response or output variable) on one (or more) explanatory (input) variables. Once identified, this relationship between the variables can be formally represented as a linear/additive function or equation. As is the case with many other modeling techniques, regression aims to capture the functional relationship between and among the characteristics of the real world and describe this relationship with a mathematical model, which may then be used to discover and understand the complexities of reality—explore and explain relationships or forecast future occurrences.

Regression can be used for two different purposes: hypothesis testing—investigating potential relationships between different variables—and prediction/forecasting—estimating values of response variables based on one or more explanatory variables. These two uses are not mutually exclusive. The explanatory power of regression is also the foundation of its prediction ability. In hypothesis testing (theory building), regression analysis can reveal the existence and strength and the directions of relationships between a number of explanatory variables (often represented with x_i) and the response variable (often represented with y). In prediction, regression identifies additive mathematical relationships (in the form of an equation) between one or more explanatory variables and a response variable. Once determined, this equation can be used to forecast the values of response variable for a given set of values of the explanatory variables.

Correlation Versus Regression

Because regression analysis originated in correlation studies, and because both methods attempt to describe the association between two (or more) variables, these two terms are often confused by professionals and even by scientists. Correlation makes no a priori assumption as to whether one variable is dependent on the other(s) and is not concerned with the relationship between variables; instead, it gives an estimate about the degree of association between the variables. On the other hand, regression attempts to describe the dependence of a response variable on one or more explanatory variables, where it implicitly assumes that there is a one-way causal effect from the explanatory variables to the response variable, regardless of whether the path of effect is direct or indirect. Also, whereas correlation is interested in low-level relationships between two variables, regression is concerned with the relationships between all explanatory variables and the response variable.

Simple Versus Multiple Regression

If a regression equation is built between one response variable and one explanatory variable, then it is called *simple regression*. For instance, the regression equation built to predict or explain the relationship between the height of a person (explanatory variable) and the weight of a person (response variable) is a good example of simple regression. Multiple regression is an extension of simple regression in which there are multiple explanatory variables. For instance, in the previous example, if we were to include not only the height of a person but also other personal characteristics (e g., BMI, gender, ethnicity) to predict the weight of the person, then we would be performing multiple regression analysis. In both cases, the relationships between the response variables and the explanatory variables are linear and additive in nature. If the relationships are not linear, then we might want to use one of many other nonlinear regression methods to better capture the relationships between the input and output variables.

How to Develop a Linear Regression Model

To understand the relationship between two variables, the simplest thing to do is to draw a scatter plot, where the y-axis represents the values of the response variable and the x-axis represents the values of the explanatory variable (see Figure 5.11). Such a scatter plot would show the changes in the response variable as a function of the changes in the explanatory variable. In the example shown in Figure 5.11, there seems to be a positive relationship between the two: As the explanatory variable values increase, so does the response variable.

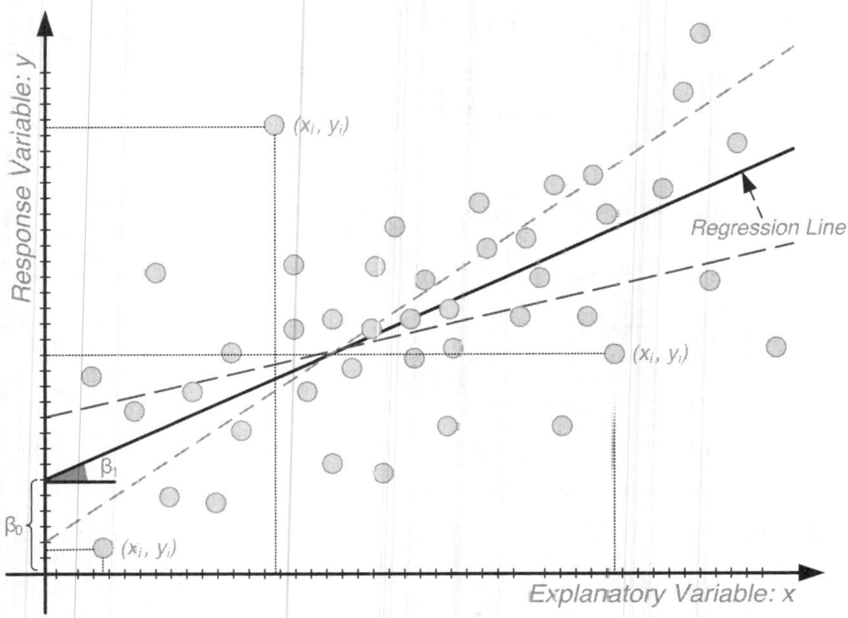

Figure 5.11 A Scatter Plot with a Simple Linear Regression Line

Simple regression analysis aims to find a mathematical representation of this relationship. In reality, it tries to find the signature of a straight line passing right in between the plotted dots (representing the observation/historical data) in such a way that it minimizes the distance between the dots and the line (the predicted values on the theoretical regression line). Several methods/algorithms have been proposed to identify the regression line, and the one that is most commonly used is called the ordinary least squares (OLS) method. The OLS method aims to minimize the sum of squared residuals (i.e., squared vertical distances between the observation and the regression point) and leads to a mathematical expression for the estimated value of the regression line (known as the β parameter). For simple linear regression, the aforementioned relationship between the response variable (y) and the explanatory variable(s) (x) can be shown as a simple equation, as follows:

$$y = \beta_0 + \beta_1 x$$

In this equation, β_0 is called the intercept, and β_1 is called the slope. Once OLS determines the values of these two coefficients, the simple equation can be used to forecast the values of y for given values of x. The sign and the value of β_1 also reveal the direction and the strength of the relationship between the two variables.

If the model is of a multiple linear regression type, then more coefficients need to be determined—one for each additional explanatory variable. As the following formula shows, the additional explanatory variable would be multiplied with new β_i coefficients and summed together to establish a linear additive representation of the response variable.

$$y = \beta_0 + \beta_1 x_1 + \beta_2 x_2 + \ldots + \beta_n x_n$$

How to Tell if a Model Is Good Enough

For a variety of reasons, sometimes models do not do a good job representing reality. Regardless of the number of explanatory variables included, there is always a possibility of not having a good model, and therefore a linear regression model needs to be assessed for its fit (the degree to which it represents the response variable). In the simplest sense, a well-fitting regression model results in predicted values close to the observed data values. For the numeric assessment, three statistical measures are often used in evaluating the fit of a regression model. R^2 (R-squared), the overall F-test, and the root mean square error (RMSE). All three of these measures are based on sums of square errors (how far the data is from the mean and how far the data is from the model's predicted values). Different combinations of these two values provide different information about how the regression model compares to the mean model.

Of the three, R^2 has the most useful and understandable meaning because of its intuitive scale. The value of R^2 ranges from 0 to 1 (corresponding to the amount of variability, expressed as a percentage), with 0 indicating that the relationship and the prediction power of the proposed model is not good and 1 indicating that the proposed model is a perfect fit that produces exact predictions (which is almost never the case). Good R^2 values usually come close to 1, and the closeness is a matter of the phenomenon being modeled; for example, while an R^2 value of 0.3 for a linear regression model in social science may be considered good enough, an R^2 value of 0.7 in engineering might not be considered a good fit. Improvement in the regression model can be achieved by adding more explanatory variables, taking some of the variables out of the model, or using different data transformation techniques, which would result in comparative increases in an R^2 value.

Figure 5.12 shows the process flow involved in developing regression models. As shown in the figure, the model development task is followed by model assessment. Model assessment involves assessing the fit of the model and also, because of restrictive assumptions with which linear models have to comply, it also involves examining the validity of the model.

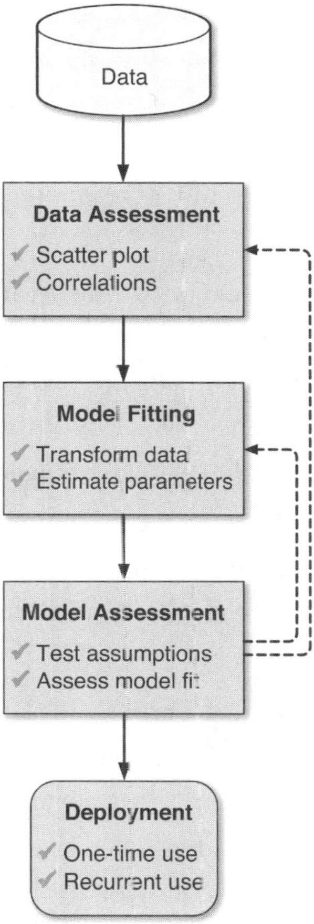

Figure 5.12 Process Flow for Developing Regression Models

The Most Important Assumptions in Linear Regression

Even though they are still the top choice of many data analysis (both for explanatory as well as predictive modeling purposes), linear regression models suffer from several highly restrictive assumptions. The validity of a linear model depends on its ability to comply with these assumptions. Here are the most common assumptions:

- **Linearity.** This assumption states that the relationship between the response variable and the explanatory variables are linear. That is, the expected value of the response variable is a straight-line function of each explanatory variable, while holding all other explanatory variables fixed. Also, the slope of the line does not depend on the values of the other variables. This implies that the effects of different explanatory variables on the expected value of the response variable are additive in nature.

- **Independence (of errors).** This assumption states that the errors of the response variable are uncorrelated with each other. This independence of the errors is weaker than actual statistical independence, which is a stronger condition and is often not needed for linear regression analysis.
- **Normality (of errors).** This assumption states that the errors of the response variable are normally distributed. That is, they are supposed to be totally random and should not represent any nonrandom patterns.
- **Constant variance (of errors).** This assumption, also called *homoscedasticity*, states that the response variables must have the same variance in their error, regardless of the values of the explanatory variables. In practice, this assumption is invalid if the response variable varies over a wide enough range or scale.
- **Multicollinearity.** This assumption states that the explanatory variables are not correlated (i.e., do not replicate the same but provide a different perspective of the information needed for the model). Multicollinearity can be triggered by having two or more perfectly correlated explanatory variables present in the model (e.g., if the same explanatory variable is mistakenly included in the model twice, one with a slight transformation of the same variable). A correlation-based data assessment usually catches this error.

Statistical techniques have been developed to identify the violation of these assumptions, and several techniques have been created to mitigate them. A data modeler needs to be aware of the existence of these assumptions and put in place a way to assess models to make sure they are compliant with the assumptions they are built on.

Logistic Regression

Logistic regression is a very popular, statistically sound, probability-based classification algorithm that employs supervised learning. It was developed in the 1940s as a complement to linear regression and linear discriminant analysis methods. It has been used extensively in numerous disciplines, including the medical and social science fields. Logistic regression is similar to linear regression in that it also aims to regress to a mathematical function that explains the relationship between the response variable and the explanatory variables, using a sample of past observations (training data). It differs from linear regression on one major point: Its output (response variable) is a class as opposed to a numeric variable. That is, whereas linear regression is used to estimate a continuous numeric variable, logistic regression is used to classify a categorical variable. Even though the original form of logistic regression was developed for a binary output variable (e.g., 1/0, yes/no, pass/fail, accept/reject), the present-day modified version is capable of predicting multiple-class output variables (i.e., multinomial logistic regression). If there is only one predictor variable and one predicted variable, the method is called simple logistic regression; this is similar to using the term simple linear regression for a linear regression model with only one independent variable.

In predictive analytics, logistic regression models are used to develop probabilistic models between one or more explanatory or predictor variables (which may be a mix of both continuous and categorical variables) and a class or response variable (which may be binomial/binary or multinomial/multiple-class variables). Unlike ordinary linear regression, logistic regression is used for predicting categorical (often a binary) outcomes of the response variable—treating the response variable as the outcome of a Bernoulli trial. Therefore, logistic regression takes the natural logarithm of the odds of the response variable to create a continuous criterion as a transformed version of the response variable. Thus the logit transformation is referred to as the *link function* in logistic regression; even though the response variable in logistic regression is categorical or binomial, the logit is the continuous criterion on which linear regression is conducted.

Figure 5.13 shows a logistic regression function where the odds are represented on the x-axis (i.e., a linear function of the independent variables), and the probabilistic outcome is shown on the y-axis (i.e., response variable values change between 0 and 1).

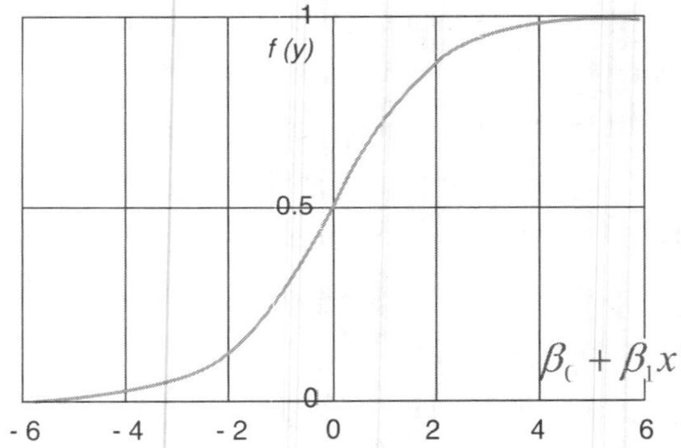

Figure 5.13 The Logistic Function

The logistic function, which is $f(y)$ in Figure 5.13, is the core of logistic regression, which can only take values between 0 and 1. The following equation is a simple mathematical representation of this function:

$$f(y) = \frac{1}{1 + e^{-(\beta_0 + \beta_1 x)}}$$

The logistic regression coefficients (the βs) are usually estimated using the maximum-likelihood estimation method. Unlike with linear regression with normally distributed residuals, it is not possible to find

a closed-form … izes the
likelihood fu… ad. This
process begi… ises the
parameters s… repeats
this iterative … l or the
improvement … s said to
have complet…

DATA MINING ALGORITHMS
NEAREST NEIGHBOUR - INSTANCE BASED LEARNING & FUNCTION IS APPROXIMATED LOCALLY K'S NEAREST NEIGHBOUR (K-NN ALGORITHM)
(REGRESSION) - AVERAGING THE VALUES OF THE NEAREST NEIGHBOURS
SIMILARITY MEASURE: DISTANCE METRIC
- DEFINING A METRIC FOR MEASURING THE CASES
① PARAMETER SELECTION
② CROSS-VALIDATION
ARTIFICIAL NEURAL NETWORKS - NON-PARAMETRIC, LEARNS FROM DATA, CAN GENERALIZE.
① BIOLOGICAL & ARTIFICIAL NEURAL NETWORKS - IMPLEMENTED AS SOFTWARE SIMULATIONS OF THE MASSIVELY PARALLEL PROCESSES INVOLVED IN PROCESSING INTERCONNECTED ELEMENTS
COMPONENTS OF ANN:
1. PROCESSING ELEMENTS - ARTIFICIAL NEURONS
2. NETWORK STRUCTURE - COLLECTION OF NEURONS GROUPED INTO LAYERS
ANN - BLACK BOXES: CAN SOLVE COMPLEX PROBLEMS BUT LACKING EXPLANATIONS FOR THEIR CAPABILITIES
SUPPORT VECTOR MACHINES
PROCESS:
1. PRE-PROCESS THE DATA (CLEAN DATA)
2. DEVELOP THE MODEL (OPTIMAL COMBINATION OF PARAMETERS)
ANN
- HEURISTIC PATH
- APPLICATIONS & EXPERIMENTING PRECEDING THEORY
SVM
- SOUND STATISTICAL THEORY, IMPLEMENTATION, EXPERIMENTS
LINEAR REGRESSION •
· CORRELATION VS REGRESSION
· SIMPLE VS MULTIPLE REGRESSION
ASSUMPTIONS IN LR
· LINEARITY · MULTICOLLINEARITY
· INDEPENDENCE
· NORMALITY
· CONSTANT VARIANCE
• LOGISTIC REGRESSION
TIME-SERIES FORECASTING

Time-Se…

Sometim… …le) may not have distinctly identifiable explanatory variables, or there may be too many of them in a highly complex relationship. In such cases, if the data is available in the desired format, a prediction model, called a *time-series forecast*, can be developed. A time series is a sequence of data points of the variable of interest, measured and represented at successive points in time and spaced at uniform time intervals. Examples of time series include monthly rain volumes in a geographic area, the daily closing value of the stock market indices, and daily sales totals for a grocery store. Often, time series are visualized using a line chart. Figure 5.14 shows an example of a time series of sales volumes for the years 2008 through 2012, on a quarterly basis.

Time-series forecasting involves using a mathematical model to predict future values of the variable of interest, based on previously observed values. Time-series plots or charts look very similar to simple linear regression scatter plots in that there are two variables: the response variable and the time variable. Beyond this similarity, there is hardly any other commonality between the two. While regression analysis is often used in testing theories to see if current values of one or more explanatory variables explain (and hence predict) the response variable, time-series models are focused on extrapolating the time-varying behavior to estimate future values.

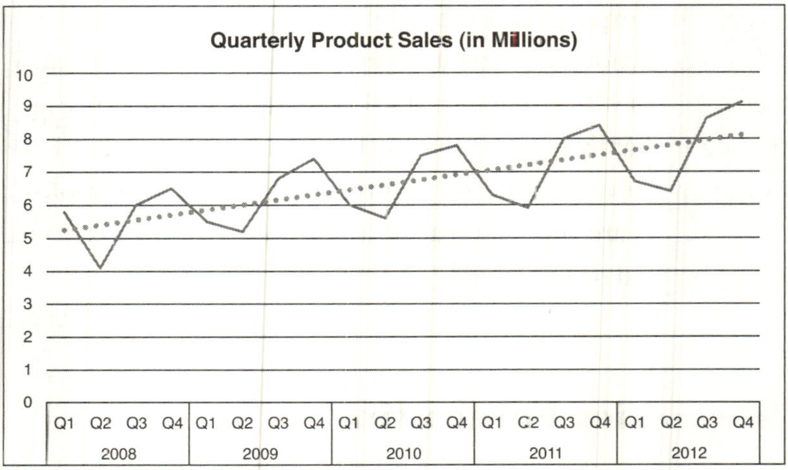

Figure 5.14 Sample Time-Series Data on Quarterly Sales Volumes

Time-series forecasting assumes that all the explanatory variables are aggregated and consumed in the response variable's time-variant behavior. Therefore, capturing time-variant behavior is a way to predict the future values of the response variable. To do that, the pattern is analyzed and decomposed into its main components: random variations, time trends, and seasonal cycles. The time-series example shown in Figure 5.14 illustrates all these distinct patterns.

The techniques used to develop time-series forecasts ranges from very simple (e.g., the naïve forecast which suggests that today's forecast is the same as yesterday's actual) to very complex (e.g., ARIMA, which combines autoregressive and moving average patterns in data). The most popular techniques are perhaps the averaging methods that include simple averages, moving averages, weighted moving averages, and exponential smoothing. Many of these techniques also have advanced versions where seasonality and trend can also be taken into account for better and more accurate forecasting. The accuracy of a method is usually assessed by computing its error (i.e., calculated deviation between actuals and forecasts for the past observations) via mean absolute error (MAE), mean squared error (MSE), or mean absolute percent error (MAPE). Even though they all use the same core error measure, these three assessment methods emphasize different aspects of the error, some panelizing larger errors more so than the others.

Application Example: Data Mining for Complex Medical Procedures

Health care has become one of the most important issues related to quality of life in the United States and elsewhere around the world. While the demand for health care services is increasing because of the aging population, the supply side is having problems keeping up with the level and quality of service. In order to close the gap, health care systems need to significantly improve their operational effectiveness and efficiency. Effectiveness (doing the right thing, such as diagnosing and treating accurately) and efficiency (doing it the right way, such as using the least resources and time) are the two fundamental pillars on which the health care system can be revived. A promising way to improve health care is to take advantage of predictive modeling techniques along with large and feature-rich data sources (true reflections of medical and health care experiences) to support accurate and timely decision making.

According to the American Heart Association, cardiovascular disease (CVD) is the underlying cause of more than 20% of deaths in the United States. Since 1900, CVD has been the number-one killer every year except 1918, which was the year of the great flu pandemic. CVD kills more people than the next four leading causes of death—cancer, chronic lower respiratory disease, accidents, and diabetes mellitus—combined. More than half of all CVD deaths can be attributed to coronary diseases. Not only does CVD take a huge toll on the personal health and well-being of the population, it is also a great drain on the health care resources in the United States and elsewhere in the world. The direct and indirect costs associated with CVD for a year are estimated to exceed $500 billion. A common surgical procedure to cure a large variant of CVD is called coronary artery bypass grafting (CABG). The cost of a CABG surgery depends on the patient and factors related to the service provider, but the average rate in the United States is between $50,000 and $100,000. As an illustrative example, Delen et al. (2012) carried out an analytics study in which they used various predictive modeling methods to predict the outcome of a CABG surgery, and they applied an information fusion–based sensitivity analysis on the trained models to better understand the importance of the prognostic factors. The main goal was to illustrate that predictive and explanatory analysis of large and feature-rich data sets provides invaluable information for making more efficient and effective decisions in health care.

Research Method

Figure 5.15 shows the model development and testing process Delen et al. used. They employed four different types of prediction models (artificial neural networks, support vector machines, and two types of decision trees, C5 and CART) and went through a large number of experimental runs to calibrate the modeling parameters for each model type. Once the models were developed, then they used the models on the text data set. Finally, the trained models were exposed to a sensitivity analysis procedure to measure the contributions of the variables. Table 5.2 shows the test results for the four different types of prediction models.

Results

In this study, Delen et al. showed the power of data mining in predicting the outcome and in analyzing the prognostic factors of complex medical procedures such as CABG surgery. They showed that using a number of prediction methods (as opposed to only one) in a competitive experimental setting has the potential to produce better predictive as well as explanatory results. Among the four methods they used, SVMs produced the best results, with prediction accuracy of 88% on the test data sample. The information fusion–based sensitivity analysis revealed the ranked importance of the independent variables. Some of the top variables identified in this analysis overlapped with the most important variables identified in previously conducted clinical and biological studies; this confirms the validity and effectiveness of the proposed data mining methodology.

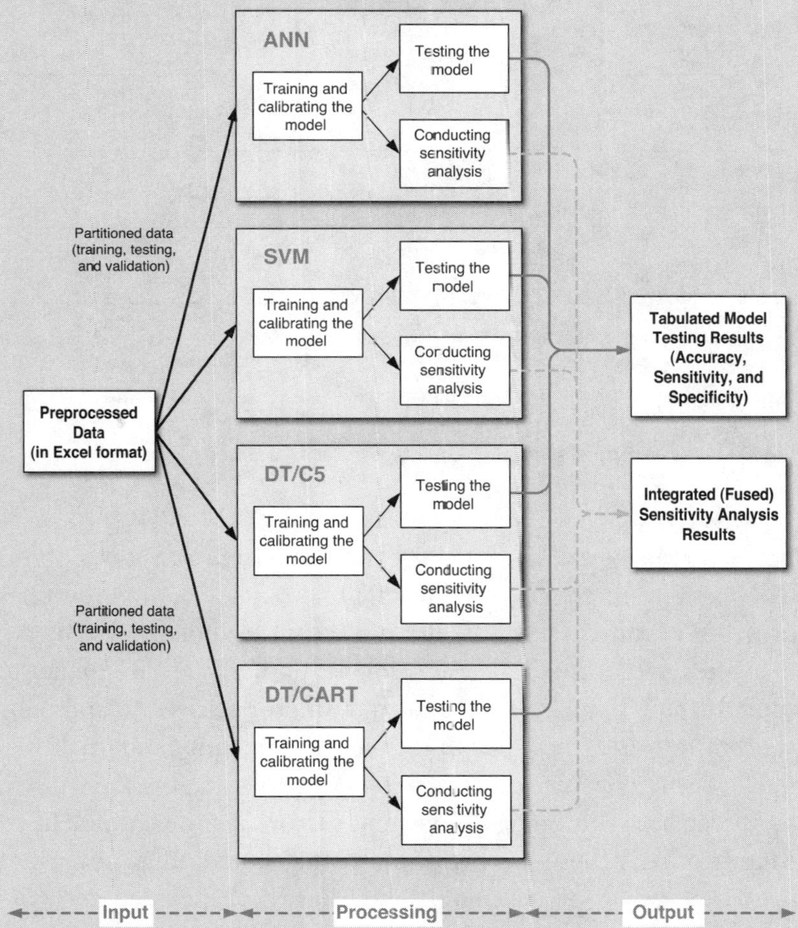

Figure 5.15 A Process Map for Training and Testing of the Four Predictive Models

Table 5.2 Prediction Accuracy Results for All Four Model Types, Based on the Test Data Set

Model Type*		Confusion Matrices**		Accuracy†	Sensitivity†	Specificity†
		Pos (1)	**Neg (0)**			
ANN	Pos (1)	749	230	74.72%	76.51%	72.93%
	Neg (0)	265	714			
SVM	Pos (1)	876	103	87.74%	89.48%	86.01%
	Neg (0)	137	842			
C5	Pos (1)	876	103	79.62%	80.29%	78.96%
	Neg (0)	137	842			
CART	Pos (1)	660	319	71.15%	67.42%	74.87%
	Neg (0)	246	733			

* Acronyms for model types: ANN: artificial neural network; SVM: support vector machine; C5: a popular decision tree algorithm; CART: classification and regression trees.

** Prediction results for the test data samples are shown in a confusion matrix, where the rows represent the actuals, and columns represent the predicted cases.

† Accuracy, sensitivity, and specificity are the three performance measures that were used in comparing the four prediction models.

From a managerial standpoint, clinical decision support systems that use the outcomes of data mining studies (such as the ones presented in this case study) are not meant to replace health care managers and/or medical professionals. Rather, they are meant to support them in making accurate and timely decisions to optimally allocate resources in order to increase the quantity and quality of medical services. We still have a long way to go before we will see these decision aids being used extensively in health care practices. Among others, there are behavioral, ethical, and political reasons for the resistance to adopting these tools. Perhaps need and government incentives for better health care systems will expedite their adoption.

Source: Delen, D., A. Oztekin, & L. Tomak, "An analytic approach to better understanding and management of coronary surgeries," *Decision Support Systems*, 52(3) (2012): 698–705.

References

Aizerman, M., E. Braverman, & L. Rozonoer. (1964). "Theoretical Foundations of the Potential Function Method in Pattern Recognition Learning," *Automation and Remote Control*, 25: 821–837.

Collins, E., S. Ghosh, & C. L. Scofield. (1988). "An Application of a Multiple Neural Network Learning System to Emulation of Mortgage Underwriting Judgments." *IEEE International Conference on Neural Networks*, 2: 459–466.

Das, R., I. Turkoglu, & A. Sengur. (2009). "Effective Diagnosis of Heart Disease Through Neural Networks Ensembles," *Expert Systems with Applications*, 36: 7675–7680.

Delen, D., A. Oztekin, & L. Tomak. (2012). "An analytic approach to better understanding and management of coronary surgeries," *Decision Support Systems*, 52(3): 698–705.

Delen, D., R. Sharda, & M. Bessonov. (2006). "Identifying Significant Predictors of Injury Severity in Traffic Accidents Using a Series of Artificial Neural Networks," *Accident Analysis and Prevention*, 38(3): 434–444.

Delen, D., & E. Sirakaya. (2006). "Determining the Efficacy of Data-Mining Methods in Predicting Gaming Ballot Outcomes," *Journal of Hospitality & Tourism Research*, 30(3): 313–332.

Fadlalla, A., & C. Lin. (2001). "An Analysis of the Applications of Neural Networks in Finance," *Interfaces*, 31(4).

Haykin, S. S. (2009). *Neural Networks and Learning Machines*, 3rd ed. Upper Saddle River, NJ: Prentice Hall.

Hill, T., T. Marquez, M. O'Connor, & M. Remus. (1994). "Neural Network Models for Forecasting and Decision Making," *International Journal of Forecasting*, 10.

Hopfield, J. (1982). "Neural Networks and Physical Systems with Emergent Collective Computational Abilities," *Proceedings of the National Academy of Science*, 79(8).

Liang, T. P. (1992). "A Composite Approach to Automated Knowledge Acquisition," *Management Science*, 38(1).

Loeffelholz, B., E. Bednar, & K. W. Bauer. (2009). "Predicting NBA Games Using Neural Networks," *Journal of Quantitative Analysis in Sports*, 5(1).

McCulloch, W. S., & W. H. Pitts. (1943). "A Logical Calculus of the Ideas Imminent in Nervous Activity," *Bulletin of Mathematical Biophysics*, 5.

Minsky, M., & S. Papert. (1969). *Perceptrons*. Cambridge, MA: MIT Press.

Piatesky-Shapiro, G. *ISR: Microsoft Success Using Neural Network for Direct Marketing*, http://kdnuggets.com/news/94/n9.txt (accessed May 2009).

Principe, J. C., N. R. Euliano, & W. C. Lefebvre. (2000). *Neural and Adaptive Systems: Fundamentals Through Simulations*. New York: Wiley.

Sirakaya, E., D. Delen, & H-S. Choi. (2005). "Forecasting Gaming Referenda," *Annals of Tourism Research*, 32(1): 127–149.

Tang, Z., C. de Almieda, & P. Fishwick. (1991). "Time-Series Forecasting Using Neural Networks vs. Box-Jenkins Methodology," *Simulation*, 57(5).

Walczak, S., W. E. Pofahi, & R. J. Scorpio. (2002). "A Decision Support Tool for Allocating Hospital Bed Resources and Determining Required Acuity of Care," *Decision Support Systems*, 34(4).

Wallace, M. P. (2008, July). "Neural Networks and Their Applications in Finance," *Business Intelligence Journal*, 67–76.

Wen, U-P., K-M. Lan, & H-S. Shih. (2009). "A Review of Hopfield Neural Networks for Solving Mathematical Programming Problems," *European Journal of Operational Research*, 198: 675–687.

Wilson, C. I., & L. Threapleton. (2003, May 17–22). "Application of Artificial Intelligence for Predicting Beer Flavours from Chemical Analysis," *Proceedings of the 29th European Brewery Congress*, Dublin, Ireland. http://neurosolutions.com/resources/apps/beer.html (accessed May 2009).

Wilson, R., & R. Sharda. (1994). "Bankruptcy Prediction Using Neural Networks," *Decision Support Systems*, 11.

Zahedi, F. (1993). *Intelligent Systems for Business: Expert Systems with Neural Networks*. Belmont, CA: Wadsworth.

6

Text Analytics and Sentiment Analysis

The information age that we are living in is characterized by rapid growth in the amount of data and information collected, stored, and made available in electronic format. A vast majority of business data is stored in text documents that are virtually unstructured. According to a study by Merrill Lynch and Gartner, 85% of all corporate data is captured and stored in some sort of unstructured form (McKnight, 2005). The same study also stated that this unstructured data is doubling in size every 18 months. Because knowledge is power in today's business world, and knowledge is derived from data and information, businesses that effectively and efficiently tap into their text data sources will have the necessary knowledge to make better decisions, leading to a competitive advantage over those businesses that lag behind. This is where the need for text analytics and text mining fits into the big picture of today's businesses.

Even though the overarching goal for both text analytics and text mining is to turn unstructured textual data into actionable information through the application of natural language processing (NLP) and analytics, their definitions are somewhat different, at least to some experts in the field. According to many, *text analytics* is a broader concept that includes information retrieval (e.g., searching and identifying relevant documents for a given set of key terms) as well as information extraction, data mining, and web mining, whereas *text mining* is primarily focused on discovering new and useful knowledge from textual data sources.

Figure 6.1 illustrates the relationships between text analytics and text mining, along with other related application areas. The bottom of the figure lists the main disciplines (the foundation of the house) that play critical roles in the development of these increasingly more popular application areas.

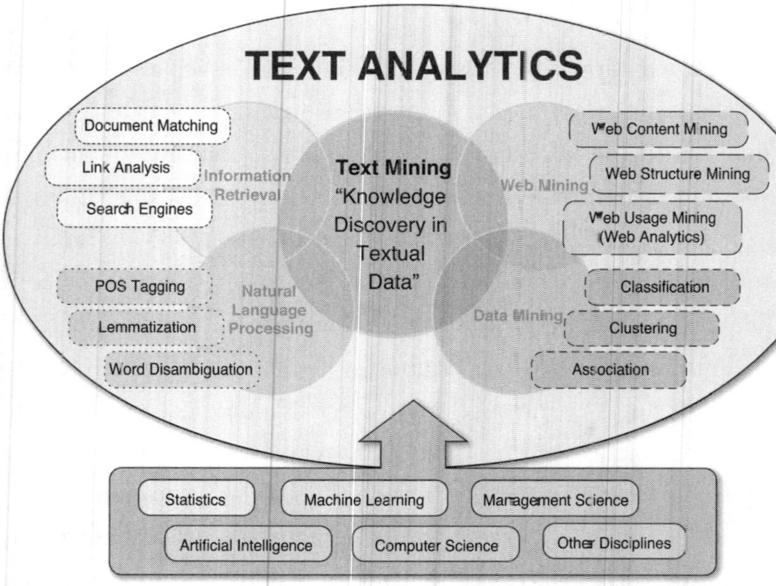

Figure 6.1 Text Mining and Enabling Disciplines

Compared to text mining, text analytics is a relatively new term. With the recent emphasis on analytics, many related technical application areas (e.g., consumer analytics, completive analytics, visual analytics, social analytics) have jumped on the analytics bandwagon. While the term *text analytics* is more commonly used in business application context, text mining is frequently used in academic research circles. Even though these terms may be defined somewhat differently at times, text analytics and text mining are usually used synonymously.

Text mining (also known as text data mining or knowledge discovery in textual databases) is the semiautomated process of extracting patterns (i.e., useful information and knowledge) from large amounts of unstructured data. Remember that data mining is the process of identifying valid, novel, potentially useful, and ultimately understandable patterns in data stored in structured databases, where the data are organized in records structured by categorical, ordinal, or continuous variables. Text mining is the same as data mining in that it has the same purpose and uses the same processes, but with text mining the input to the process is a collection of unstructured (or less structured) data files, such as Word documents, PDF files, text excerpts, XML files, and so on. In essence, text mining can be thought of as a process (with two main steps) that starts with imposing structure on the text-based data sources followed by extracting relevant information and knowledge from this structured text-based data by using data mining techniques and tools.

The benefits of text mining are obvious in areas where very large amounts of textual data are being generated, such as law (e.g., court orders), academic research (e.g., research articles), finance (e.g., quarterly reports), medicine (e.g., discharge summaries), biology (e.g., molecular interactions), technology (e.g., patent files), and marketing (e.g., customer comments). For example, free-form text-based interactions with customers in the form of complaints (or praises) and warranty claims can be used to objectively identify product and service characteristics that are deemed to be less than perfect and can be used as input to improve product development and service allocations. Likewise with market outreach programs and focus groups that generate large amounts of data. By not restricting product or service feedback to a codified form, customers can present, in their own words, what they think about a company's products and services.

Another area where the automated processing of unstructured text has had a lot of impact is in email and other electronic communications. Text mining not only can be used to classify and filter junk email, but it can also be used to automatically prioritize email based on importance level as well as generate automatic responses (Weng

& Liu, 2004). The following are some of the most popular application areas of text mining:

- **Information extraction.** Text mining can identify key phrases and relationships within text by looking for predefined objects and sequences in text by using pattern matching. Perhaps the most commonly used form of information extraction is entity extraction. Named entity extraction includes named entity recognition (recognition of known entity names—for people and organizations, place names, temporal expressions, and certain types of numeric expressions, using existing knowledge of the domain), co-reference resolution (detection of co-reference and anaphoric links between text entities), and relationship extraction (identification of relationships between entities).
- **Topic tracking.** Based on a user profile and documents that a user views, text mining can predict other documents that might be of interest to the user.
- **Summarization.** Text mining can be used to summarize a document to save the reader time.
- **Categorization.** Text mining can be used to identify the main themes of a document and then place the document into a predefined set of categories, based on those themes.
- **Clustering.** Text mining can be used to group similar documents without having a predefined set of categories.
- **Concept linking.** Text mining can be used to connect related documents by identifying their shared concepts and, by doing so, help users find information that they perhaps would not find by using traditional search methods.
- **Question answering.** Text mining can be used to find the best answer to a given question through knowledge-driven pattern matching.

Text mining has its own language, with many technical terms and acronyms. Following is a list of some of the terms and concepts that are commonly used in the context of text mining:

- **Unstructured data (versus structured data).** Structured data has a predetermined format. It is usually organized into records with simple data values (categorical, ordinal, and continuous variables) and stored in databases. In contrast, unstructured data does not have a predetermined format and is stored in the form of textual documents. In essence, the structured data is for the computers to process, while the unstructured data is for humans to process and understand.
- **Corpus.** In linguistics, a corpus (plural corpora) is a large and structured set of texts (now usually stored and processed electronically), prepared for the purpose of conducting knowledge discovery.
- **Terms.** A term is a single-word or multiword phrase extracted directly from the corpus of a specific domain by means of natural language processing (NLP) methods.
- **Concepts.** Concepts are features generated from a collection of documents by means of manual, statistical, rule-based, or hybrid categorization methodologies. Concepts result from higher-level abstraction than do terms.
- **Stemming.** Stemming is the process of reducing inflected words to their stem (or base or root) form. For instance, *stemmer*, *stemming*, and *stemmed* are all based on the root *stem*.
- **Stop words.** Stop words (or noise words) are words that are filtered out prior to or after processing of natural language data (i.e., text). Even though there is no universally accepted list of stop words, most natural language processing tools use a list that includes articles (a, am, the, of, etc.), auxiliary verbs (is, are, was, were, etc.), and context-specific words that are deemed not to have differentiating value.

- **Synonyms and polysemes.** Synonyms are syntactically different words (i.e., spelled differently) with identical or at least similar meanings (e.g., movie, film, and motion picture). In contrast, polysemes, also called homonyms, are syntactically identical words (i.e., spelled exactly the same) but have different meanings (e.g., bow can mean "to bend forward," "the front of a ship," "a weapon that shoots arrows," or "a kind of tied ribbon").
- **Tokenizing.** A token is a categorized block of text in a sentence. The block of text corresponding to the token is categorized according to the function it performs. This assignment of meaning to blocks of text is known as tokenizing. A token can look like anything; it just needs to be a useful part of the structured text.
- **Term dictionary.** A term dictionary is a collection of terms specific to a narrow field that can be used to restrict the extracted terms within a corpus.
- **Word frequency.** The number of times a word is found in a specific document is called the word frequency.
- **Part-of-speech tagging.** This is the process of marking up the words in a text that correspond to a particular part of speech (e.g., nouns, verbs, adjectives, adverbs), based on each word's definition and the context in which it is used.
- **Morphology.** Morphology is a branch of the field of linguistics and a part of natural language processing that studies the internal structure of words (patterns of word formation within a language or across languages).
- **Term-by-document matrix (occurrence matrix).** This type of matrix is a common representation schema for the frequency-based relationship between terms and documents in tabular format, where terms are listed in rows, documents are listed in columns, and the frequency between the terms and documents is listed in cells as integer values.

- **Singular-value decomposition (latent semantic indexing).** This dimensionality reduction method is used to transform a term-by-document matrix to a manageable size by generating an intermediate representation of the frequencies using a matrix manipulation method similar to principal-component analysis.

Natural Language Processing

Some of the early text mining applications used a simplified representation called bag-of-words when introducing structure to a collection of text-based documents in order to classify them into two or more predetermined classes or to cluster them into natural groupings. In the bag-of-words model, text—such as a sentence, paragraph, or complete document—is represented as a collection of words, disregarding the grammar or the order in which the words appear. The bag-of-words model is still used in some simple document classification tools. For instance, in spam filtering, an email message can be modeled as an unordered collection of words (a bag of words) that is compared against two different predetermined bags. One bag is filled with words found in spam messages and the other is filled with words found in legitimate emails. Although some of the words are likely to be found in both bags, the "spam" bag will contain spam-related words such as *stock market*, *Viagra*, and *buy* much more frequently than the legitimate bag, which will contain more words related to the user's friends or workplace. The level of match between a specific email's bag of words and the two bags containing the descriptors determines the characterization of the email as either spam or legitimate.

Naturally, we humans do not use words without some order or structure. We use words in sentences, which have semantic as well as syntactic structure. Thus, automated techniques (e.g., text mining) need to look for ways to go beyond the bag-of-words interpretation and incorporate more and more semantic structure into their operations. The current trend in text mining is toward including many of the advanced features that can be obtained using natural language processing.

It has been shown that the bag-of-words method may not produce good enough information content for text mining tasks (e.g., classification, clustering, association). A good example of this can be found in evidence-based medicine. A critical component of evidence-based medicine is incorporating the best available research findings into the clinical decision-making process, which involves appraisal of the information collected from the printed media for validity and relevance. Several researchers from University of Maryland developed evidence assessment models using a bag-of-words method (Lin & Demner-Fushman, 2005). They employed popular machine learning methods along with more than half a million research articles collected from MEDLINE (Medical Literature Analysis and Retrieval System). In their models, they represented each abstract as a bag of words, where each stemmed term represented a feature. Despite using popular classification methods with proven experimental design methodologies, their prediction results were not much better than simple guessing, which may indicate that the bag-of-words method is not generating a good enough representation of the research articles in this domain; hence, more advanced techniques such as natural language processing are needed.

Natural language processing (NLP) is an important component of text mining and is a subfield of artificial intelligence and computational linguistics. It studies the problem of understanding natural human language, with the view of converting depictions of human language (e.g., textual documents) into more formal representations (in the form of numeric and symbolic data) that are easier for computer programs to manipulate. The goal of NLP is to move beyond syntax-driven text manipulation (which is often called "word counting") to a true understanding and processing of natural language that considers grammatical and semantic constraints as well as the context.

The definition and scope of the word *understanding* is one of the major discussion topics in NLP. Considering that the natural human language is vague and that a true understanding of meaning requires extensive knowledge of a topic (beyond what is in the words, sentences, and paragraphs), will computers ever be able to understand natural language the same way and with the same accuracy that humans do? Probably not! NLP has come a long way from the days of simple word counting, but it has an even longer way to go to really understanding natural human language. The following are just a few of the challenges commonly associated with the implementation of NLP:

- **Part-of-speech tagging.** It is difficult to mark up terms in a text as corresponding to a particular part of speech (e.g., nouns, verbs, adjectives, adverbs) because the part of speech depends not only on the definition of the term but also on the context within which the term is used.
- **Text segmentation.** Some written languages, such as Chinese, Japanese, and Thai, do not have single-word boundaries. In these instances, the text-parsing task requires the identification of word boundaries, which is often difficult. Similar challenges in speech segmentation emerge when analyzing spoken language because sounds representing successive letters and words blend into each other.
- **Word sense disambiguation.** Many words have more than one meaning. Selecting the meaning that makes the most sense can only be accomplished by taking into account the context within which the word is used.
- **Syntactic ambiguity.** The grammar for natural languages is ambiguous; that is, multiple possible sentence structures often need to be considered. Choosing the most appropriate

structure usually requires a fusion of semantic and contextual information.

- **Imperfect or irregular input.** Foreign or regional accents and vocal impediments in speech and typographical or grammatical errors in texts make the processing of language an even more difficult task.
- **Speech acts.** A sentence can often be considered a request for action. The sentence structure alone may not contain enough information to define the action. For example, "Can you pass the class?" requests a simple yes/no answer, whereas "Can you pass the salt?" is a request for a physical action to be performed.

It is a longstanding dream of the artificial intelligence community to come up with algorithms that are capable of automatically reading and obtaining knowledge from text. By applying a learning algorithm to parsed text, researchers from Stanford University's NLP lab have developed methods that can automatically identify concepts and relationships between those concepts in text. By applying a unique procedure to large amounts of text, their algorithms automatically acquire hundreds of thousands of items of world knowledge and use them to produce significantly enhanced repositories for WordNet. WordNet is a laboriously hand-coded database of English words, their definitions, sets of synonyms, and various semantic relationships between synonym sets. It is a major resource for NLP applications, but it has proven to be very expensive to build and maintain manually. By automatically inducing knowledge into WordNet, the potential exists to make WordNet an even greater and more comprehensive resource for NLP at a fraction of the cost.

One prominent area where the benefits of NLP and WordNet are already being harvested is in customer relationship management (CRM). Broadly speaking, the goal of CRM is to maximize customer value by better understanding and effectively responding to their actual and perceived needs. An important area of CRM, where NLP

CHAPTER 5 OF DR. DELEN'S BOOK TALKS ABOUT THE DIFFERENT DATA MINING ALGORITHMS AVAILABLE. IT TALKED ABOUT 7 DATA MINING ALGORITHMS AND THEY ARE: NEAREST NEIGHBOUR,- INSTANCE BASED LEARNING & FUNCTION IS APPROXIMATELY LOCALLY K'S; SIMILARITY MEASURE: THE DISTANCE METRIC - DEFINING A METRIC FOR MEASURING THE CASES; ARTIFICIAL NEURAL NETWORKS- NON-PARAMETIC, LEARNS FROM DATA, & CAN GENERALIZE; SUPPORT VECTOR MACHINES - SUPERVISED LEARNING METHOD THAT PRODUCE INPUT/OUTPUT FUNCTIONS FROM A SET OF LABELED TRAINING DATA; LINEAR REGRESSION - DEPENDENCE OF A VARIABLE (RESPONSE OR OUTPUT), ON ONE (OR MORE) EXPLANATORY (INPUT) VARIABLES; LOGISTIC REGRESSION - SIMILAR TO LINEAR EXCEPT ITS OUTPUT IS A CLASS AS OPPOSED TO A NUMERIC VARIABLE; AND, TIME-SERIES FORECASTING - SEQUENCE OF DATA POINTS OF THE VARIABLE OF INTEREST, MEASURED & REPRESENTED AT SUCCESSIVE POINTS IN TIME & SPACED AT UNIFORM TIME INTERVALS.

is making a significant impact, is sentiment analysis. Sentiment analysis is a technique used to detect favorable and unfavorable opinions toward specific products and services, using large numbers of textual data sources (customer feedback in the form of Web postings). Detailed coverage of sentiment analysis is given later in this chapter.

NLP has successfully been applied to a variety of domains for a variety of tasks via computer programs to automatically process natural human language that previously could be done only by humans. Following are among the most popular of these tasks:

- **Question answering.** NLP can be used to automatically answer a question posed in natural language—that is, to produce a human-language answer to a human-language question. To find the answer to a question, the computer program may use either a prestructured database or a collection of natural language documents (a text corpus such as the World Wide Web).
- **Automatic summarization.** A computer program can be used to create a shortened version of a textual document that contains the most important points of the original document.
- **Natural language generation.** Systems can convert information from computer databases into readable human language.
- **Natural language understanding.** Systems can convert samples of human language into more formal representations that are easier for computer programs to manipulate.
- **Machine translation.** Systems can automatically translate one human language to another.
- **Foreign language reading.** A computer program can assist a nonnative language speaker in reading a foreign language with correct pronunciation and accents on different parts of the words.

- **Foreign language writing.** A computer program can assist a nonnative language user in writing in a foreign language.
- **Speech recognition.** Systems can convert spoken words to machine-readable input. Given a sound clip of a person speaking, such a system produces a text dictation.
- **Text-to-speech.** A computer program can automatically convert normal language text into human speech. This is also called speech synthesis.
- **Text proofing.** A computer program can read a proof copy of a text in order to detect and correct any errors.
- **Optical character recognition.** A system can automatically translate images of handwritten, type-written, or printed text (usually captured by a scanner) into machine-editable textual documents.

The success and popularity of text mining depends greatly on advancements in NLP in both generation as well as understanding of human languages. NLP enables the extraction of features from unstructured text so that a wide variety of data mining techniques can be used to extract knowledge (novel and useful patterns and relationships) from it. In that sense, simply put, text mining is a combination of NLP and data mining.

Text Mining Applications

As the amount of unstructured data collected by organizations increases, so do the value proposition and popularity of text mining tools. Many organizations are now realizing the importance of extracting knowledge from their document-based data repositories through the use of text mining tools. Following are only a small subset of the exemplary application categories of text mining.

Marketing Applications

Text mining can be used to increase cross-selling and up-selling by analyzing the unstructured data generated by call centers. Text generated by call-center notes as well as transcriptions of voice conversations with customers can be analyzed by text mining algorithms to extract novel, actionable information about customers' perceptions about a company's products and services. In addition, blogs, user reviews of products at independent Websites, and discussion board postings are gold mines of customer sentiments. This rich collection of information, properly analyzed, can be used to increase satisfaction and the overall lifetime value of a customer.

Text mining has become invaluable for customer relationship management. Companies can use text mining to analyze rich sets of unstructured text data, combined with relevant structured data extracted from organizational databases, to predict customer perceptions and subsequent purchasing behavior. For instance text mining can be used to meaningfully improve the ability of a mathematical model to predict customer churn (i.e., customer attrition) so that those customers identified as most likely to leave a company could be accurately identified for retention tactics.

Treating products as sets of attribute–value pairs rather than as atomic entities can potentially boost the effectiveness of many business applications, including demand forecasting, assortment optimization, product recommendations, assortment comparison across retailers and manufacturers, and product supplier selection. Ghani et al. (2006) used text mining to develop a system capable of inferring implicit and explicit attributes of products to enhance retailers' ability to analyze product databases. The proposed system allows a business to represent its products in terms of attributes and attribute values without much manual effort. The system learns these attributes by applying supervised and semi-supervised learning techniques to product descriptions found on retailers' Websites.

Security Applications

One of the largest and most prominent text mining applications in the security domain is the highly classified ECHELON surveillance system. As rumor has it, ECHELON is capable of identifying the content of telephone calls, faxes, emails, and other types of data, intercepting information sent via satellites, public switched telephone networks, and microwave links.

In 2007, EUROPOL developed an integrated system capable of accessing, storing, and analyzing vast amounts of structured and unstructured data sources in order to track transnational organized crime. Called the Overall Analysis System for Intelligence Support (OASIS), this system aims to integrate the most advanced data and text mining technologies available in today's market. The system has enabled EUROPOL to make significant progress in supporting its law enforcement objectives at the international level (EUROPOL, 2007).

The U.S. Federal Bureau of Investigation (FBI) and the Central Intelligence Agency (CIA), under the direction of the Department for Homeland Security, are jointly developing a supercomputer data and text mining system. The system is expected to create a gigantic data warehouse along with a variety of data and text mining modules to meet the knowledge-discovery needs of federal, state, and local law enforcement agencies. Prior to this project, the FBI and CIA each had its own separate databases, with little or no interconnection.

Another security-related application of text mining is in the area of deception detection. Applying text mining to a large set of real-world criminal (person-of-interest) statements, Fuller et al. (2008) developed prediction models to differentiate deceptive statements from truthful ones. Using a rich set of cues extracted from the textual statements, the model predicted the holdout samples with 70% accuracy, which is believed to be a significant success considering that the cues are extracted only from textual statements (with no verbal or visual cues). Furthermore, compared to other deception-detection techniques, such as polygraph, this method is nonintrusive and widely applicable to not only textual data but also (potentially) to transcriptions of voice recordings. A more detailed description of text-based deception detection is provided at the end of this chapter as an application example.

Biomedical Applications

Text mining holds great potential for the medical field in general and biomedicine in particular for several reasons. First, the published literature and publication outlets (especially with the advent of open source journals) in the field are expanding exponentially. Second, compared to most other fields, the medical literature is more standardized and orderly, making it a more "minable" information source. Finally, the terminology used in this literature is relatively constant, having a fairly standardized ontology. A few exemplary studies have successfully used text mining techniques in extracting novel patterns from biomedical literature.

Experimental techniques such as DNA microarray analysis, serial analysis of gene expression (SAGE), and mass spectrometry proteomics, among others, are generating large amounts of data related to genes and proteins. As in any other experimental approach, it is necessary to analyze this vast amount of data in the context of previously known information about the biological entities under study. The literature is a particularly valuable source of information for experiment validation and interpretation. Therefore, the development of automated text mining tools to assist in such interpretation is one of the main challenges in current bioinformatics research.

Knowing the location of a protein within a cell can help elucidate its role in biological processes and determine its potential as a drug target. Numerous location-prediction systems are described in the literature; some focus on specific organisms, whereas others attempt to analyze a wide range of organisms. Shatkay et al. (2007) proposed a comprehensive system that uses several types of sequence- and text-based features to predict the locations of proteins. The main novelty of their system lies in the way in which it selects its text sources and features and integrates them with sequence-based features. Shatkay et al. (2007) tested the system on previously used data sets and on new data sets devised specifically to test the predictive power of the system. The results showed that their system consistently beat previously reported results.

Chun et al. (2006) described a system that extracts disease–gene relationships from literature accessed via MEDLINE. They constructed a dictionary for disease and gene names from six public databases and extracted relationship candidates by using dictionary matching. Because dictionary matching produces a large number of false positives, Chun et al. developed a method of machine learning–based named entity recognition (NER) to filter out false recognitions of disease/gene names. They found that the success of relation extraction is heavily dependent on the performance of NER filtering and that the filtering improved the precision of relation extraction by 26.7%, at the cost of a small reduction in recall.

Figure 6.2 shows a simplified depiction of a multilevel text analysis process for discovering gene–protein relationships (or protein–protein interactions) in the biomedical literature (Nakov et al., 2005). This simplified example uses a simple sentence from biomedical text. As shown in the figure, first (at the bottom three levels), the text is tokenized using part-of-speech tagging and shallow parsing. The tokenized terms (words) are then matched (and interpreted) against the hierarchical representation of the domain ontology to derive the gene–protein relationship. Application of this method (and/or some variation of it) to the biomedical literature offers great potential for decoding the complexities in the Human Genome Project.

The issue of text mining is of great importance to publishers that hold large databases of information requiring indexing for better retrieval. This is particularly true in scientific disciplines, in which highly specific information is often contained in written text. Initiatives have been launched, such as *Nature*'s proposal for an Open Text Mining Interface (OTMI) and the National Institutes of Health's Journal Publishing Document Type Definition (DTD), that would provide semantic cues to machines to answer specific queries contained within text without removing publisher barriers to public access.

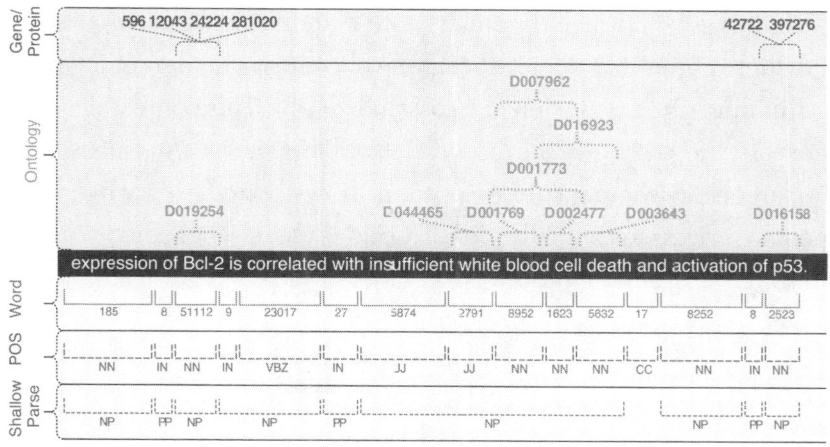

Figure 6.2 Multilevel Analysis of Text for Gene/Protein Interaction Identification

Academic institutions have also launched text mining initiatives. For example, the National Centre for Text Mining, a collaborative effort between the Universities of Manchester and Liverpool in the United Kingdom, provides customized tools, research facilities, and advice on text mining to the academic community. With an initial focus on text mining in the biological and biomedical sciences, research has since expanded into the social sciences. In the United States, the School of Information at the University of California, Berkeley, is developing a program called BioText to assist bioscience researchers in text mining and analysis.

The Text Mining Process

In order to be successful, text mining studies should follow a sound methodology based on best practices. Remember from Chapter 3, "The Data Mining Process," that CRISP-DM is the industry standard for data mining projects; likewise, a standardized process model is needed for text mining. Even though most parts of CRISP-DM are also applicable to text mining projects, a specific process model for text mining would include much more elaborate data preprocessing activities. Figure 6.3 depicts a high-level context diagram of a typical text mining process (Delen & Crossland, 2008). This context diagram presents the scope of the process, emphasizing its interfaces with the larger environment. In essence, it draws boundaries around the specific process to explicitly identify what is included in (and excluded from) the text mining process.

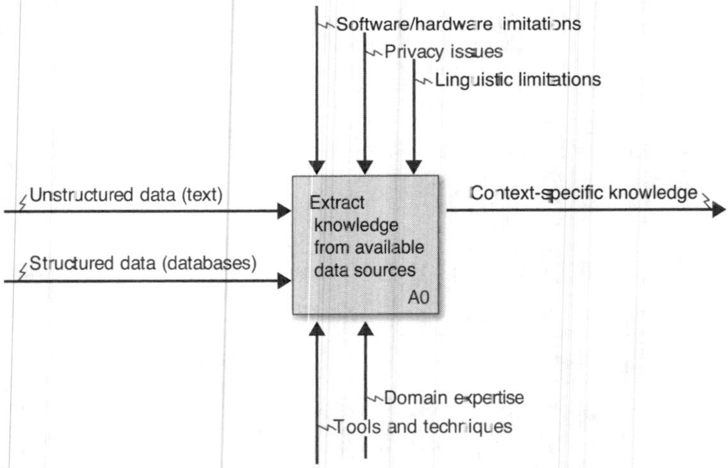

Figure 6.3 Context Diagram for the Text Mining Process

As the context diagram indicates, the input (the inward connection to the left edge of the box) into the text-based knowledge discovery process is the unstructured as well as structured data collected, stored, and made available to the process. The output (the outward extension from the right edge of the box) of the process is the context-specific knowledge that can be used for decision making. The controls, also called the constraints (the inward connection to the top edge of the box), of the process include software and hardware limitations, privacy issues, and difficulties related to processing the text that is presented in the form of natural language. The mechanisms (the inward connection to the bottom edge of the box) of the process include proper techniques, software tools, and domain expertise. The primary purpose of text mining (within the context of knowledge discovery) is to process unstructured (textual) data (along with structured data, if relevant to the problem being addressed and available) to extract meaningful and actionable patterns for better decision making.

At a very high level, the text mining process can be broken down into three consecutive tasks, each of which has specific inputs to generate certain outputs (see Figure 6.4). If, for some reason, the output of a task is not what was expected, a backward redirection to the previous task execution is necessary.

Task 1: Establish the Corpus

The main purpose of the first task in the text mining process is to collect all the documents related to the context (i.e., domain of interest) being studied. This collection may include textual documents, XML files, emails, Web pages, and short notes. In addition to the readily available textual data, voice recordings may also be transcribed using speech-recognition algorithms and made a part of the text collection.

Once collected, the text documents are transformed and organized in a manner such that they are all in the same representational form (e.g., ASCII text files) for computer processing. The organization of the documents can be as simple as a collection of digitized text excerpts stored in a file folder, or it can be a list of links to a collection of Web pages in a specific domain. Many commercially available text mining software tools could accept these as input and convert them into a flat file for processing. Alternatively, the flat file can be prepared outside the text mining software and then presented as the input to the text mining application.

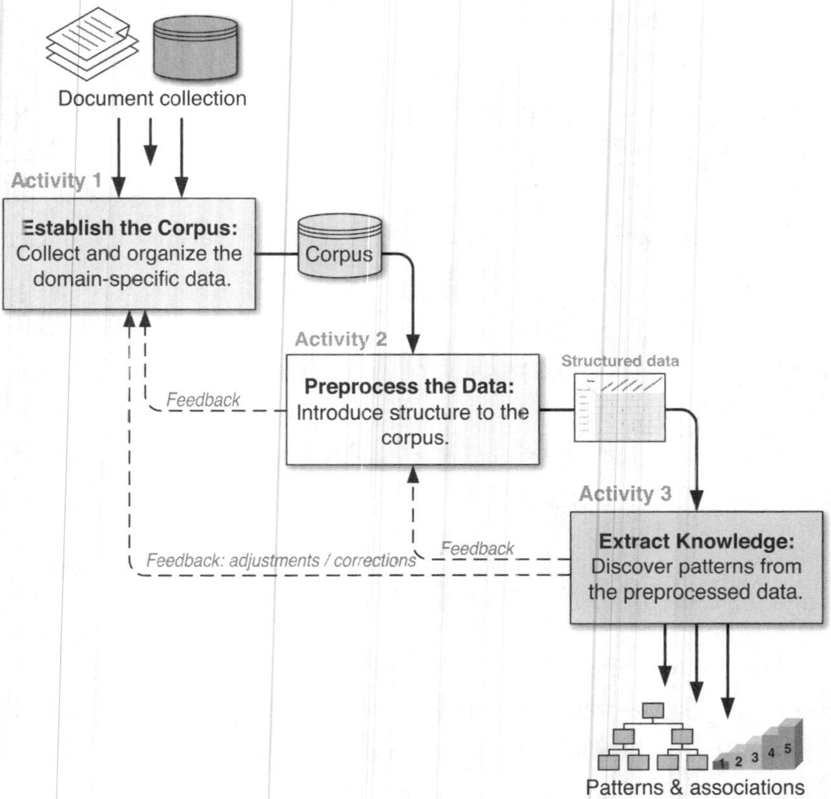

Figure 6.4 The Three-Task Text Mining Process

Task 2: Preprocess the Data: Create the Term–Document Matrix

In task 2 of the text mining process, the digitized and organized documents (the corpus) are used to create the term–document matrix (TDM). In the TDM, rows represent the documents, and columns represent the terms. The relationships between the terms and documents are characterized by indices (i.e., a relational measure that can be as simple as the number of occurrences of the term in respective documents). Figure 6.5 shows a typical example of a TDM.

Terms / Documents	Investment risk	Project management	Software engineering	Development	SAP	...	
Document 1	1			1			
Document 2		1					
Document 3			3		1		
Document 4		1					
Document 5			2	1			
Document 6	1			1			
...							

Figure 6.5 An Example of a Term–Document Matrix

The goal is to convert the list of organized documents (the corpus) into a TDM where the cells are filled with the most appropriate indices. The assumption is that the essence of a document can be represented with a list of the frequency of the terms used in that document. However, are all terms important when characterizing documents? Obviously, the answer is "no." Some terms, such as articles, auxiliary verbs, and terms used in almost all the documents in the corpus, have no differentiating power and therefore should be excluded from the indexing process. This list of terms, commonly called *stop terms* or *stop words*, is specific to the domain of study and should be identified by the domain experts. On the other hand, the experts might choose a set of predetermined terms under which the documents are to be indexed (called *include terms* or a *dictionary*). In addition, synonyms (pairs of terms that are to be treated the same) and specific phrases (e.g., "Eiffel Tower") can also be provided so that the index entries are more accurate.

Another filtration that should take place to accurately create the indices is stemming, which refers to the reduction of words to their roots so that, for example, different grammatical forms or declinations of a verb are identified and indexed as the same word. For example, stemming would ensure that *modeling* and *modeled* will be recognized as the word *model*.

The first generation of a TDM includes all the unique terms identified in the corpus (as its columns), excluding the ones in the stop term list; all the documents (as its rows); and the occurrence count of each term for each document (as its cell values). If, as is commonly the case, the corpus includes a rather large number of documents, then there is a very good chance that the TDM will have a very large number of terms. Processing such a large matrix might be time-consuming and, more importantly, might lead to extraction of inaccurate patterns. At this point, one has to decide the following: (1) What is the best way to represent the indices? and (2) How can we reduce the dimensionality of this matrix to a manageable size?

What Is the Best Way to Represent the Indices?

Once the input documents are indexed and the initial word frequencies (by document) are computed, a number of additional transformations can be performed to summarize and aggregate the extracted information. The raw term frequencies generally reflect on how salient or important a word is in each document. Specifically, words that occur with greater frequency in a document are better descriptors of the contents of that document. However, it is not reasonable to assume that the word counts themselves are proportional to their importance as descriptors of the documents. For example, if a word occurs one time in document A but three times in document B, then it is not necessarily reasonable to conclude that the word is three times as important a descriptor of document B as of document

A. In order to have a more consistent TDM for further analysis, these raw indices need to be normalized. As opposed to showing the actual frequency counts, the numeric representation between terms and documents can be normalized using a number of alternative methods. The following are a few of the most commonly used normalization methods (StatSoft, 2014):

- **Log frequencies.** The raw frequencies can be transformed using the log function. This transformation "dampens" the raw frequencies and how they affect the results of subsequent analysis.
- **Binary frequencies.** An even simpler transformation than log frequencies can be used to enumerate whether a term is used in document binary frequencies. The resulting TDM matrix will contain only 1s and 0s to indicate the presence or absence of the respective words. Again, this transformation dampens the effect of the raw frequency counts on subsequent computations and analyses.
- **Inverse document frequencies.** Another issue to consider related to the indices used in further analyses is the relative document frequencies (df) of different terms. For example, a term such as *guess* may occur frequently in all documents, whereas another term, such as *software*, may appear only a few times. The reason is that one might make *guesses* in various contexts, regardless of the specific topic, whereas *software* is a more semantically focused term that is only likely to occur in documents that deal with computer software. A common and very useful transformation that reflects both the specificity of words (i.e., document frequencies) and the overall frequencies of their occurrences (i e., term frequencies) is the so-called inverse document frequency (Manning et al. 2008).

How Can We Reduce the Dimensionality of the TDM Matrix?

Because the TDM is often very large and rather sparse (with most of the cells filled with zeros), an important question is "How can we reduce the dimensionality of this matrix to a manageable size?" Several options are available for managing matrix size:

- A domain expert can go through the list of terms and eliminate those that do not make much sense for the context of the study (this is a manual, labor-intensive process).
- We can eliminate terms with very few occurrences in very few documents.
- We can transform the matrix by using singular value decomposition.

Singular value decomposition (*SVD*), which is closely related to principal-component analysis, reduces the overall dimensionality of the input matrix (i.e., the number of input documents by the number of extracted terms) to a lower dimensional space, where each consecutive dimension represents the largest degree of variability (between words and documents) possible (Manning et al. 2008). Ideally, an analyst might identify the two or three most salient dimensions that account for most of the variability (i.e., differences) between the words and documents, thus identifying the latent semantic space that organizes the words and documents in the analysis. Once such dimensions are identified, the underlying meaning of what is contained (discussed or described) in the documents has been extracted.

Task 3: Extract the Knowledge

Using a well-structured TDM, potentially augmented with other structured data elements, novel patterns are extracted in the context of the specific problem being addressed. The main categories of knowledge extraction methods are classification, clustering, association, and trend analysis. A short description of these methods follows.

Classification

Arguably the most common knowledge discovery topic in analyzing complex data sources is the classification (or categorization) of certain objects. The task is to classify a given data instance into a predetermined set of categories (or classes). As it applies to the domain of text mining, the task is known as *text categorization*, where for a given set of categories (subjects, topics, or concepts) and a collection of text documents, the goal is to find the correct topic (subject or concept) for each document, using models developed with a training data set that includes both the documents and actual document categories. Today, automated text classification is applied in a variety of contexts, including automatic or semiautomatic (interactive) indexing of text, spam filtering, Web page categorization under hierarchical catalogs, automatic generation of metadata, and detection of genre.

Clustering

Clustering is an unsupervised process whereby objects are classified into "natural" groups called *clusters*. In categorization, a collection of preclassified training examples is used to develop a model based on the descriptive features of the classes in order to classify a new unlabeled example; on the other hand, in clustering, the problem is to group an unlabeled collection of objects (e.g., documents, customer comments, Web pages) into meaningful clusters without any prior knowledge.

Clustering is useful in a wide range of applications, from document retrieval to better Web content searching. In fact, one of the prominent applications of clustering is the analysis and navigation of very large text collections, such as Web pages. The basic underlying assumption is that relevant documents tend to be more similar to each other than to irrelevant ones. If this assumption holds, the clustering of documents based on the similarity of their content improves search effectiveness.

Association

A formal definition and detailed description of *association* was provided in Chapter 4, "Data and Methods in Data Mining." The main idea in generating association rules (or solving market-basket problems) is to identify the sets that frequently go together. In text mining, *associations* specifically refer to the direct relationships between concepts (terms) or sets of concepts. The concept set association rule, relating two frequent concept sets A and C, can be quantified by the two basic measures support and confidence. In this case, *confidence* is the percentage of documents that include all the concepts in C within the same subset of those documents that include all the concepts in A. *Support* is the percentage (or number) of documents that include all the concepts in A and C. For instance, in a document collection, the concept *software implementation failure* may appear most often in association with *enterprise resource planning* and *customer relationship management*, with significant support (4%) and confidence (55%), meaning that 4% of the documents had all three concepts represented together in the same document, and of the documents that included *software implementation failure*, 55% of them also included *enterprise resource planning* and *customer relationship management*. Text mining with association rules was used to analyze published literature (news and academic articles posted on the Web) to chart the outbreak and progress of bird flu (Mahgoub et al., 2008). The idea was to automatically identify the associations among the geographic areas, spread across species, and countermeasures (treatments).

Trend Analysis

Recent methods of trend analysis in text mining have been based on the notion that the various types of concept distributions are functions of document collections; that is, different collections lead to different concept distributions for the same set of concepts. It is therefore possible to compare two distributions that are otherwise identical except that they are from different subcollections. One notable direction of this type of analyses is having two collections from the same source (e.g., from the same set of academic journals) but from different points in time. Delen and Crossland (2008) applied trend analysis to a large number of academic articles published in the three highest-rated academic journals to identify the evolution of key concepts in the field of information systems. Following is a description of this text mining application.

Application Example: Text Mining of Research Literature

Researchers conducting searches and reviews of relevant literature face an increasingly complex and voluminous task. In extending the body of relevant knowledge, it has always been important to work hard to gather, organize, analyze, and assimilate existing information from the literature, particularly from one's home discipline. With the increasing abundance of potentially significant research being reported in related fields, and even in what are traditionally deemed to be nonrelated fields of study, the researcher's task is ever more daunting, if a thorough job is desired.

In new streams of research, the researcher's task may be even more tedious and complex. Trying to ferret out relevant work that others have reported may be difficult, at best, and perhaps even nearly impossible if traditional, largely manual reviews of published literature are required. Even with a legion of dedicated graduate students or helpful colleagues, trying to cover all potentially relevant published work is problematic.

Many scholarly conferences take place every year. In addition to extending the body of knowledge of the current focus of a conference, organizers often desire to offer additional mini-tracks and workshops. In many cases, these additional events are intended to introduce the attendees to significant streams of research in related fields of study and to try to identify the "next big thing" in terms of research interests and focus. Identifying reasonable candidate topics for such mini-tracks and workshops is often subjective rather than derived objectively from the existing and emerging research.

In their research, Delen and Crossland (2008) proposed a method to greatly assist and enhance the efforts of the researchers by enabling a semiautomated analysis of large volumes of published literature through the application of text mining. Using standard digital libraries and online publication search engines, the authors downloaded and collected all the available articles for the three major journals in the field of management information systems: *MIS Quarterly* (*MISQ*), *Information Systems Research* (*ISR*), and the *Journal of Management Information Systems* (*JMIS*). In order to maintain the same time interval for all three journals (for potential comparative longitudinal studies), the journal with the most recent starting date for its digital publication availability was used as the start time for this study (i.e., *JMIS* articles have been digitally available since 1994). For each article, Delen and Crossland extracted the title, abstract, author list, published keywords, volume, issue number, and year of publication. They then loaded all the article data into a simple database file. Also included in the combined data set was a field that designated the journal type of each article for likely discriminatory analysis. Editorial notes, research notes, and executive overviews were omitted from the collection. Figure 6.6 shows how the data was presented in a tabular format.

	A	B	C	D
1	ID	YEAR	JOURNAL	ABSTRACT
2	PID001	2005	MISQ	The need for continual value innovation is driving supply chair
3	PID002	1999	ISR	Although much contemporary thought considers advanced inf
4	PID003	2001	JMIS	When producers of goods (or services) are confronted by a situ
5	PID004	1995	ISR	Preservation of organizational memory becomes increasingly
6	PID005	1994	ISR	The research reported here is an adaptation of a model develc
7	PID006	1995	MISQ	This study evaluates the extent to which the added value to cu
8	PID007	2003	MISQ	This paper reports the results(-) of a field-study of six medical
9	PID008	1999	JMIS	Researchers and managers are beginning to realize that the fu
10	PID009	2000	JMIS	The Internet commerce technologies have significantly reduce
11	PID010	1997	ISR	Adaptive Structuration Theory (AST) is rapidly becoming an inf
12	PID011	1995	JMIS	Research shows that group support systems (GSS) have drama
13	PID012	2000	MISQ	Increasingly, business leaders are demanding that IT play the r
14	PID013	2001	ISR	Alignment between business strategy and IS strategy is widely
15	PID014	1999	JMIS	A framework is outlined that includes the planning of and sett

Figure 6.6 A Sample of the Textual Data Set

In the analysis phase, Delen and Crossland chose to use only the abstract of an article as the source of information extraction. They chose not to include the keywords listed with the publications for two main reasons: (1) Under normal circumstances, the abstract would already include the listed keywords, and therefore inclusion of the listed keywords for the analysis would mean repeating the same information and potentially giving them unmerited weight; and (2) the listed keywords might be terms that authors would like their article to be associated with (as opposed to what is really contained in the article), therefore potentially introducing unquantifiable bias to the analysis of the content.

The first exploratory study was to look at the longitudinal perspective of the three journals (i.e., evolution of research topics over time). In order to conduct a longitudinal study, they divided the 12-year period (from 1994 to 2005) into four 3-year periods for each of the three journals. This framework led to 12 text mining experiments with 12 mutually exclusive data sets. At this point, for each of the 12 data sets, Delen and Crossland used text mining to extract the most descriptive terms from these collections of articles represented by their abstracts. The results were tabulated and examined for time-varying changes in the terms published in these three journals.

As a second exploration, using the complete data set (including all three journals and all four periods), Delen and Crossland conducted a clustering analysis. Clustering is arguably the most commonly used text mining technique. Clustering was used in this study to identify the natural groupings of the articles (by putting them into separate clusters) and then to list the most descriptive terms that characterized those clusters. The authors used singular-value decomposition to reduce the dimensionality of the term–document matrix and then an expectation-maximization algorithm to create the clusters. They conducted several experiments to identify the optimal number of clusters, which turned out to be nine. After the construction of the nine clusters, Delen and Crossland analyzed the content of those clusters from two perspectives: (1) representation of the journal type (see Figure 6.7) and (2) representation of time (see Figure 6.8). The idea was to explore the potential differences and/or commonalities among the three journals and potential changes in the emphasis on those clusters; that is, they wanted to answer questions such as "Are there clusters that represent different research themes specific to a single journal?" and "Is there a time-varying characterization of those clusters?" They discovered and discussed several interesting patterns using tabular and graphical representation of their findings (for further information, see Delen & Crossland, 2008).

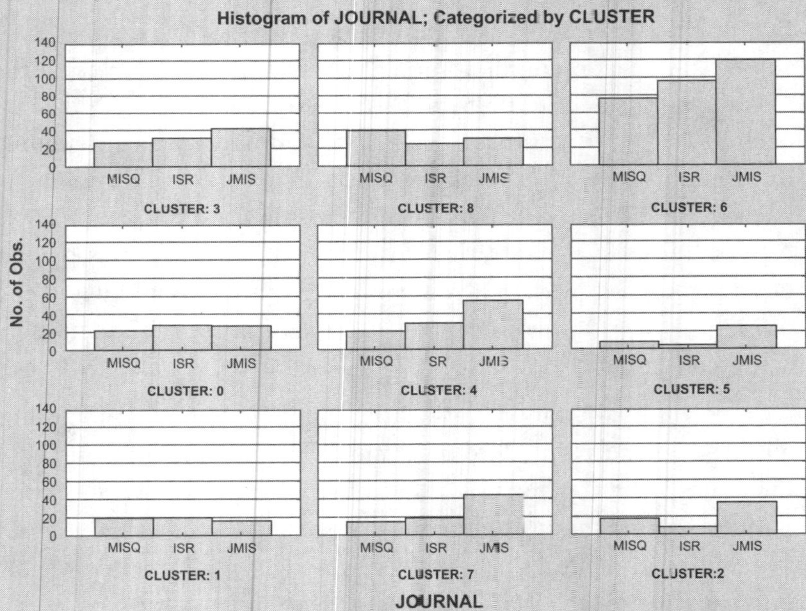

Figure 6.7 Distribution of the Number of Articles for Each Journal over the Nine Clusters

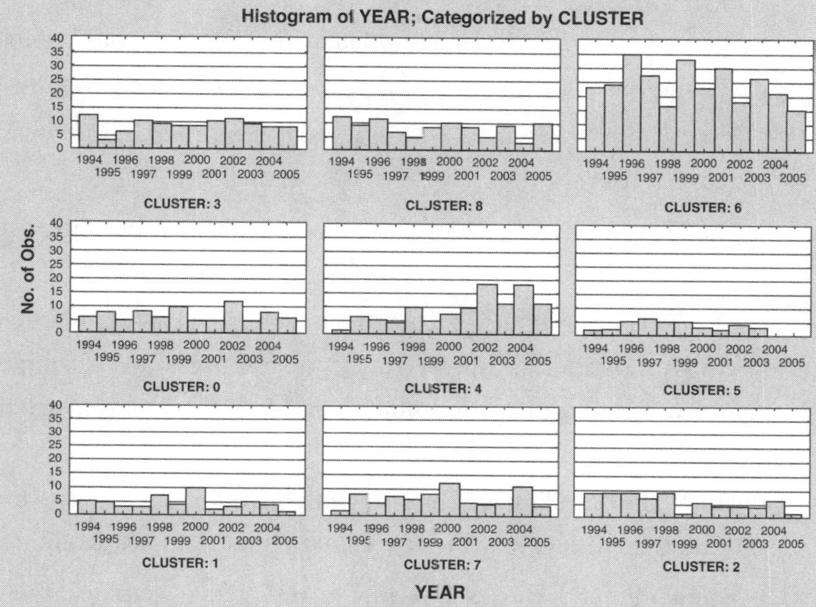

Figure 6.8 Time-Dependent Changes in the Number of Articles Published in Each Cluster

Source: Delen & Crossland, 2008.

Text Mining Tools

As more and more organizations are realizing the value of text mining, the number of software tools offered by software companies and nonprofits is also increasing. The following sections describe some of the popular text mining tools, which we classify as commercial software tools and free (and/or open source) software tools.

Commercial Software Tools

The following are some of the most popular software tools used for text mining. Note that many of these companies offer demonstration versions of their products on their Websites:

- ClearForest offers text analysis and visualization tools.
- IBM offers SPSS Modeler and Data and Text Analytics Toolkit.
- Megaputer Text Analyst offers semantic analysis of free-form text, summarization, clustering, navigation, and natural language retrieval with search-dynamic refocusing.
- SAS Text Miner provides a rich suite of text processing and analysis tools.
- KXEN Text Coder (KTC) is text analytics solution for automatically preparing and transforming unstructured text attributes into a structured representation for use in KXEN Analytic Framework.
- The Statistica Text Mining engine provides easy-to-use text mining functionally with exceptional visualization capabilities.
- VantagePoint provides a variety of interactive graphical views and analysis tools with powerful capabilities to discover knowledge from text databases.
- The WordStat analysis module from Provalis Research analyzes textual information such as responses to open-ended questions and interviews.
- Clarabridge text mining software provides an end-to-end solution for customer experience professionals wishing to transform customer feedback for marketing, service, and product improvements.

Free Software Tools

Free software tools, some of which are open source, are available from a number of nonprofit organizations:

- RapidMiner has a graphically appealing user interface and is one of the most popular free, open source software tools for data mining and text mining.

- Open Calais is an open source toolkit for including semantic functionality within your blog, content management system, Website, or application.
- GATE is a leading open source toolkit for text mining. It has a free open source framework (or SDK) and graphical development environment.
- LingPipe is a suite of Java libraries for linguistic analysis of human language.
- S-EM (Spy-EM) is a text classification system that learns from positive and unlabeled examples.
- Vivisimo/Clusty is a Web search and text-clustering engine.

Often, innovative application of text mining comes from the collective use of several software tools.

Sentiment Analysis

We humans are social beings. We are adept at utilizing a variety of means to exchange ideas. We often consult financial discussion forums before making an investment decision, we ask our friends for their opinions on a newly opened restaurant or a newly released movie, and we conduct Internet searches and read consumer reviews and expert reports before making a big purchase like a house, a car, or an appliance. We rely on others' opinions to make better decisions, especially in an area where we don't have a lot of knowledge or experience. Thanks to the growing availability and popularity of opinion-rich Internet resources such as social media outlets (e.g., Twitter, Facebook), online review sites, and personal blogs, it is now easier than ever to find opinions of others (thousands of them, as a matter of fact) on everything from the latest gadgets to political and public figures. Even though not everybody expresses opinions on the Internet, due to fast-growing numbers and capabilities of social communication channels, the numbers are increasing exponentially.

Sentiment is a difficult word to exactly define. It is often linked to or confused with other terms, like *belief*, *view*, *opinion*, and *conviction*. *Sentiment* suggests a subtle opinion reflective of one's feelings (Mejova, 2009). Sentiment has some unique properties that sets it apart from other concepts that we might want to identify in text. Often we want to categorize text by topic, which may involve dealing with whole taxonomies of topics. Sentiment classification, on the other hand, usually deals with two classes (positive versus negative), a range of polarity (e.g., star ratings for movies), or even a range in strength of opinion (Pang & Lee, 2008). These classes span many topics, users, and documents. Although dealing with only a few classes may seem like an easier task than standard text analysis, it is far from the reality.

As a field of research, sentiment analysis is closely related to computational linguistics, natural language processing, and text mining. Sentiment analysis has many names. It's often referred to as *opinion mining*, *subjectivity analysis*, and *appraisal extraction*, with some connections to *affective computing* (computer recognition and expression of emotion). The sudden upsurge of interest and activity in the area of sentiment analysis, which deals with the automatic extraction of opinions, feelings, and subjectivity in text, is creating opportunities and threats for business and individuals alike. The ones who embrace and take advantage of it will greatly benefit from it. Every opinion put on the Internet by an individual or a company will be accredited back to the originator (good or bad) and will be retrieved and mined by others (often automatically by computer programs).

Sentiment analysis tries to answer the question "What do people feel about a certain topic?" by digging into opinions of many, using a variety of automated tools. Bringing together researchers and practitioners in business, computer science, computational linguistics, data mining, text mining, psychology, and even sociology, sentiment analysis aims to expand traditional fact-based text analysis to new frontiers to realize opinion-oriented information systems. In a business setting, especially in marketing and customer relationship management, sentiment analysis seeks to detect favorable and unfavorable opinions toward specific products and/or services, using a large numbers of textual data sources (e.g., customer feedback in the form of Web postings, tweets, blogs, etc.).

Sentiment that appears in text comes in two flavors: explicit, where the text directly expresses an opinion ("It's a wonderful day"), and implicit, where the text implies an opinion ("The handle breaks too easily"). Most of the earlier work done in sentiment analysis focused on the first kind of sentiment, since it is was easier to analyze. Current trends are to implement analytical methods to consider both implicit and explicit sentiments. Sentiment polarity is a particular feature of text that sentiment analysis primarily focuses on. It is usually dichotomized into two areas—positive and negative—but polarity can also be thought of as a range. A document containing several opinionated statements would have a mixed polarity overall, which is different from not having a polarity at all (being objective) (Mejova, 2009).

Sentiment Analysis Applications

Traditional sentiment analysis methods were survey based or focus group centered; they were costly and time-consuming and, hence, were driven from a small sample of participants. In contrast, the new face of text analytics–based sentiment analysis is a limit breaker. Current solutions automate very large-scale data collection, filtering, classification, and clustering methods via natural language processing and data mining technologies that handle both factual and subjective information. Sentiment analysis is perhaps the most popular application of text analytics, tapping into data sources like tweets, Facebook posts, online communities, discussion boards, blogs, product reviews, call center logs and recordings, product rating sites, chat rooms, price comparison portals, search engine logs, and newsgroups. The following sections discuss applications of sentiment analysis to illustrate the power and the widespread coverage of this technology.

Voice of the Customer (VOC)

VOC is an integral part of an analytic CRM and customer experience management systems. As the enabler of VOC, sentiment analysis can access a company's product and service reviews (either continuously or periodically) to better understand and better manage customer complaints and praises. For instance, a motion picture advertising/marketing company might detect negative sentiments toward a movie that is about to open in theaters (based on its trailers) and quickly change the composition of the trailers and its advertising strategy (on all media outlets) to mitigate the negative impact. Similarly, a software company may detect negative buzz regarding the bugs found in its newly released product early enough to release patches and quick fixes to alleviate the situation.

Often, the focus of VOC is individual customers and their service- and support-related needs, wants, and issues. VOC draws data from the full set of customer touchpoints, including emails, surveys, call center notes/recordings, and social media postings, and it matches customer voices to transactions (e.g., inquiries, purchases, returns) and individual customer profiles captured in enterprise operational systems. VOC, mostly driven by sentiment analysis, is a key element of customer experience management initiatives, where the goal is to create an intimate relationship with the customer.

Voice of the Market (VOM)

VOM is about understanding aggregate opinions and trends. It's about knowing what stakeholders—customers, potential customers, influencers, whoever—are saying about your, and your competitors', products and services. A well-done VOM analysis helps companies with competitive intelligence and product development and positioning.

Voice of the Employee (VOE)

Traditionally, VOE has been limited to employee satisfaction surveys. Text analytics in general and sentiment analysis in specific are huge enablers of assessing VOE. Using rich, opinionated textual data is an effective and efficient way to listen to what employees are saying. As we all know, happy employees empower customer experience efforts and improve customer satisfaction.

Brand Management

Brand management focuses on listening to social media, where anyone (e.g., past/current/prospective customers, industry experts, other authorities) can post opinions that can damage or boost a reputation. A number of relatively newly launched startup companies offer analytics-driven brand management services. Brand management is product and company focused rather than customer focused. It attempts to shape perceptions rather than to manage experiences, using sentiment analysis techniques.

Financial Markets

Predicting the future values of individual (or a group of) stocks has been an interesting and seemingly unsolvable problem. What makes a stock (or a group of stocks) move up or down is anything but an exact science. Many believe that the stock market is mostly sentiment driven, making it anything but rational (especially for short-term stock movements). Therefore, use of sentiment analysis in financial markets has gained significant popularity. Automated analysis of market sentiments using social media, news, blogs, and discussion groups seems to be a proper way to compute market movements. If done correctly, sentiment analysis can identify short-term stock movements based on the buzz in the market, potentially impacting liquidity and trading.

Politics

As we all known, opinions matter a great deal in politics. Because political discussions are dominated by quotations, sarcasm, and complex references to persons, organizations, and ideas, politics is one of the most difficult—and potentially fruitful—areas for sentiment analysis. By analyzing the sentiment on election forums, it may be possible to predict who is more likely to win or lose. Sentiment analysis can help understand what voters are thinking and can clarify a candidate's position on issues. Sentiment analysis can help political organizations, campaigns, and news analysts better understand which issues and positions matter the most to voters. The technology was successfully applied by both parties to the 2008 and 2012 U.S. presidential election campaigns.

Government Intelligence

Government intelligence is another application of sentiment analysis. For example, it has been suggested that one could monitor sources for increases in hostile or negative communications. Sentiment analysis can allow the automatic analysis of the opinions that people submit about pending policy or government-regulation proposals. Furthermore, monitoring communications for spikes in negative sentiment may be of use to agencies like the Department of Homeland Security.

Other Interesting Areas

Sentiments of customers can be used to better design ecommerce sites (e.g., product suggestions, up-sell/cross-sell advertising), better place advertisements (e.g., placing dynamic advertisement of products and services that consider the sentiment on the page the user is browsing), and mine opinion- or review-oriented search engines (i.e., opinion-aggregation Websites, as alternatives to sites like Epinions, to summarize user reviews). Sentiment analysis can help with email filtration by categorizing and prioritizing incoming emails (e.g., it can detect strongly negative or flaming emails and forward them to a proper folder), as well as citation analysis, where it can determine whether an author is citing a piece of work as supporting evidence or as research that he or she dismisses.

The Sentiment Analysis Process

Because of the complexity of the problem of sentiment analysis (e.g., underlying concepts, expressions in text, the context in which text is expressed), there is not a readily available standardized process for conducting sentiment analysis. However, based on the published work in the field of sentiment analysis so far (both on research methods and range of applications), a multistep, simple, logical process, as shown in Figure 6.9, seems to be an appropriate methodology for sentiment analysis. These logical steps are iterative (i.e., feedbacks, corrections, and iterations are part of the discovery process) and experimental in nature, and once completed and combined, capable of producing desired insight about the opinions in the text collection.

Step 1: Sentiment Detection

After the retrieval and preparation of the text documents, the first main task in sensitivity analysis is the detection of objectivity. Here the goal is to differentiate between a fact and an opinion, which may be viewed as classification of text as objective or subjective. This may also be characterized as calculation of O-S polarity (objectivity–subjectivity polarity, which may be represented with a numeric value ranging from 0 to 1). If the objectivity value is close to 1, then there is no opinion to mine (i.e., it is a fact), and therefore the process goes back and grabs the next text data to analyze. Usually opinion detection is based on the examination of adjectives in text. For example, the polarity of "what a wonderful work" can be determined relatively easily by looking at the adjective.

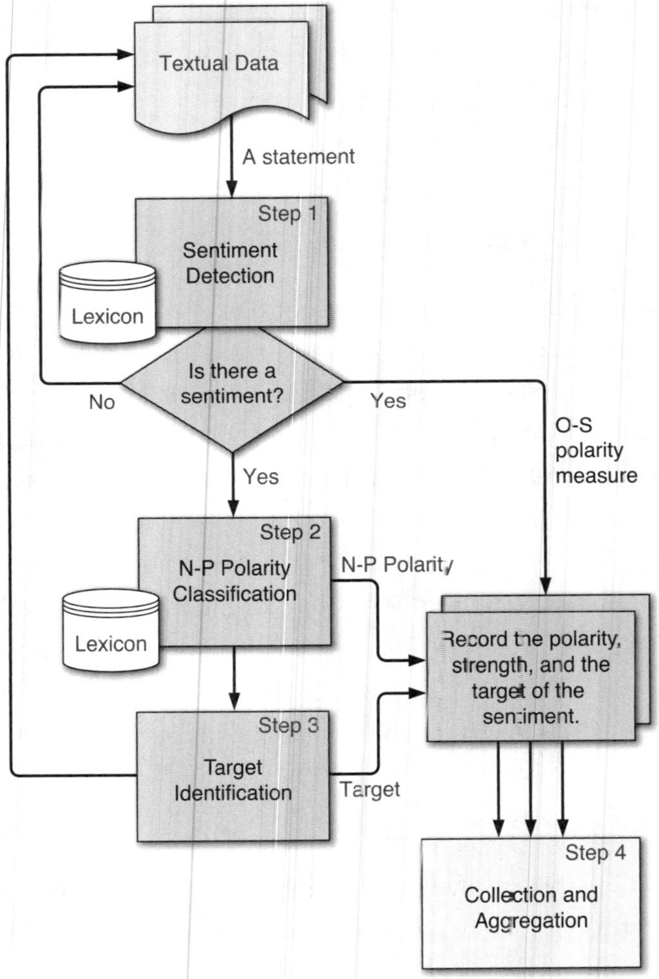

Figure 6.9 A Multistep Process for Sentiment Analysis

Step 2: N-P Polarity Classification

The second main task is polarity classification. Given an opinionated piece of text, the goal is to classify the opinion as falling under one of two opposing sentiment polarities or locate its position on the continuum between these two polarities (Pang & Lee, 2008). When viewed as a binary feature, polarity classification is the binary classification task of labeling an opinionated document as expressing either an overall positive opinion or an overall negative opinion (e.g., thumbs up/thumbs down). In addition to the identification of N-P polarity, one should also be interested in identifying the strength of the sentiment; as opposed to just positive, it may be expressed as mildly, moderately, strongly, or very strongly positive. A lot of research has been done on product or movie reviews where the definitions of "positive" and "negative" are quite clear. Other tasks, such as classifying news as "good" or "bad," present some difficulty. For instance, an article may contain negative news without explicitly using any subjective words or terms. Furthermore, these classes usually appear intermixed when a document expresses both positive and negative sentiments. Then the task can be to identify the main (or dominating) sentiment of the document. Still, for lengthy texts, the tasks of classification may need to be done at several levels: term, phrase, sentence, and perhaps document level. For those, it is common to use the outputs of one level as the inputs for the next higher layer.

Step 3: Target Identification

The goal of this step is to accurately identify the target of the expressed sentiment (e.g., a person, a product, an event). The difficulty of this task depends largely on the domain of the analysis. Even though it is usually easy to accurately identify the target for product or movie reviews, because the review is directly connected to the target, it may be quite challenging in other domains. For instance, lengthy, general-purpose text such as Web pages, news articles, and blogs do not always have predefined topics that they are assigned to, and they often mention many objects, any of which may be deduced as the target. Sometimes there is more than one target in a sentiment sentence, which is the case in comparative texts. A subjective comparative sentence orders objects in order of preferences, such as "This laptop computer is better than my desktop PC." Such sentences can be identified using comparative adjectives and adverbs (e.g., *more*, *less*, *better*, *longer*), superlative adjectives (e.g., *most*, *least*, *best*) and other words, such as *same*, *differ*, *win*, *prefer*, etc. Once the sentences have been retrieved, the objects can be put in an order that is most representative of their merits, as described in text.

Step 4: Collection and Aggregation

Once the sentiments of all text data points in the document are identified and calculated, in this step, they are aggregated and converted to a single sentiment measure for the whole document. This aggregation may be as simple as summing up the polarities and strengths of all texts or as complex as using semantic aggregation techniques from natural language processing to come up with the ultimate sentiment.

The Methods Used for Polarity Identification

As mentioned in the previous section, polarity identification can be made at the word, term, sentence, or document level. The most granular level for polarity identification is the word level. Once the polarity identification is made at the word level, it can be aggregated to the next higher level, and then the next until the level of aggregation desired from the sentiment analysis is reached. There seem to be two dominant techniques used for identification of polarity at the word/term level, each of which has advantages and disadvantages:

- Using a lexicon as a reference library—either developed manually or automatically, by an individual for a specific task or developed by an institution for general use
- Using a collection of training documents as the source of knowledge about the polarity of terms within a specific domain

(i.e., inducing predictive models from opinionated textual documents)

Application Example: Text-Based Deception Detection

Driven by advancements in Web-based information technologies and increasing globalization, computer-mediated communication continues to filter into everyday life, bringing with it new venues for deception. The volume of text-based chat, instant messaging, text messaging, and text generated by online communities of practice is increasing rapidly. Even email continues to grow in use. With the massive growth of text-based communication, the potential for people to deceive others through computer-mediated communication has also grown, and such deception can have disastrous results.

Unfortunately, in general, humans tend to perform poorly at deception-detection tasks. This phenomenon is exacerbated in text-based communications. A large part of the research on deception detection (also known as *credibility assessment*) has involved face-to-face meetings and interviews. Yet, with the growth of text-based communication, text-based deception-detection techniques are essential.

Techniques for successfully detecting deception—that is, lies—have wide applicability. Law enforcement can use decision support tools and techniques to investigate crimes, conduct security screening in airports, and monitor communications of suspected terrorists. Human resources professionals might use deception-detection tools to screen applicants. These tools and techniques also have the potential to screen emails to uncover fraud or other wrongdoings committed by corporate officers. Although some people believe that they can readily identify people not being truthful, a summary of deception research showed that, on average, people are only 54% accurate in making veracity determinations (Bond & DePaulo, 2006). This figure may actually be worse when humans try to detect deception in text.

Using a combination of text mining and data mining techniques, Fuller et al. (2008) analyzed person-of-interest statements completed by people involved in crimes on military bases. In these statements, suspects and witnesses are required to write their recollection of the event in their own words. Military law enforcement personnel searched archival data for statements that they could conclusively identify as being truthful or deceptive. These decisions were made on the basis of corroborating evidence and case resolution. Once the statements were labeled as truthful or deceptive, the law enforcement personnel removed identifying information and gave the statements to the research team. In total, 371 usable statements were received for analysis. The text-based deception-detection method used by Fuller et al. (2008) was based on a process known as *message feature mining*, which relies on elements of data and text mining techniques. A simplified depiction of the process is provided in Figure 6.10.

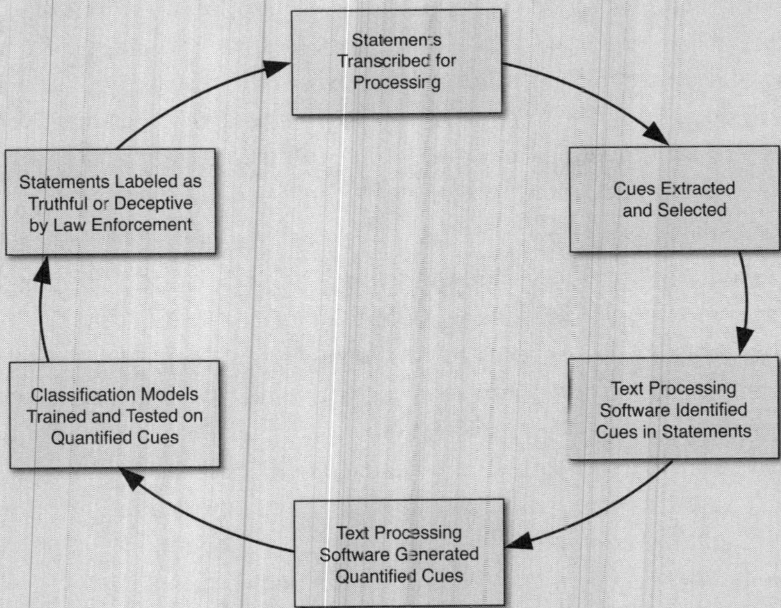

Figure 6.10 Text-Based Deception Detection Process

First, the researchers prepared the data for processing. The original handwritten statements had to be transcribed into a word processing file. Second, features (i.e., cues) were identified. The researchers identified 31 features representing categories or types of language that are relatively independent of the text content and that can be readily analyzed by automated means. For example, first-person pronouns such as *I* or *me* can be identified without analysis of the surrounding text. Table 6.1 lists the categories and a list of cues used in this study.

Table 6.1 Categories and Examples of Linguistic Features Used in Deception Detection

Number	Category	Examples of Cues
1	Quantity	Verb count, noun-phrase count, etc.
	Complexity	Average number of clauses, average sentence length, etc.
	Uncertainty	Modifiers, modal verbs, etc.
	Nonimmediacy	Passive voice, objectification, etc.
	Expressivity	Emotiveness
	Diversity	Lexical diversity, redundancy, etc.
	Informality	Typographical error ratio
	Specificity	Spatiotemporal information, perceptual information, etc.
	Affect	Positive affect, negative affect, etc.

The features were extracted from the textual statements and input into a flat file for further processing. Using several feature-selection methods along with 10-fold cross-validation, the researchers compared the prediction accuracy of three popular data mining methods. Their results indicated that neural network models performed the best, with 73.46% prediction accuracy on test data samples; decision trees performed second best, with 71.60% accuracy; and logistic regression was last, with 67.28% accuracy.

The results indicate that automated text-based deception detection has the potential to aid those who must try to detect lies in text and can be successfully applied to real-world data. The accuracy of these techniques exceeded the accuracy of most other deception-detection techniques, even though it was limited to textual cues.

Sources: C. M. Fuller, D. Biros, & D. Delen, "Exploration of Feature Selection and Advanced Classification Models for High-Stakes Deception Detection," *Proceedings of the 41st Annual Hawaii International Conference on System Sciences (HICSS)*, Big Island, HI, IEEE Press, pp. 80–99, 2008; and C. F. Bond & B. M. DePaulo, "Accuracy of Deception Judgments," *Personality and Social Psychology Reports,* 10(3): 214–234, 2006.

References

Bond C. F., & B. M. DePaulo. (2006). "Accuracy of Deception Judgments," *Personality and Social Psychology Reports,* 10(3): 214–234.

Chun, H. W., Y. Tsuruoka, J. D. Kim, R. Shiba, N. Nagata, T. Hishiki, & Jun'ichi Tsujii. (2006, January). "Extraction of Gene-Disease Relations from Medline Using Domain Dictionaries and Machine Learning," *Pacific Symposium on Biocomputing*, 11: 4–15.

Delen, D., & M. Crossland. (2008). "Seeding the Survey and Analysis of Research Literature with Text Mining," *Expert Systems with Applications*, 34(3): 1707–1720.

Etzioni, O. (1996). "The World Wide Web: Quagmire or Gold Mine?" *Communications of the ACM*, 39(11): 65–68.

EUROPOL. (2007). *EUROPOL Work Program 2007*, state-watch.org/news/2006/apr/europol-work-programme-2007.pdf (accessed October 2008).

Feldman, R., & J. Sanger. (2007). *The Text Mining Handbook: Advanced Approaches in Analyzing Unstructured Data*. Boston: ABS Ventures.

Fuller, C. M., D. Biros, & D. Delen. (2008). "Exploration of Feature Selection and Advanced Classification Models for High-Stakes Deception Detection," *Proceedings of the 41st Annual Hawaii International Conference on System Sciences (HICSS)*, Big Island, HI: IEEE Press, pp. 80–99.

Ghani, R., K. Probst, Y. Liu, M. Krema, & A. Fano. (2006). "Text Mining for Product Attribute Extraction," *SIGKDD Explorations*, 8(1): 41–48.

Grimes, S. (2011, February 17). "Seven Breakthrough Sentiment Analysis Scenarios," *InformationWeek*.

Kanayama, H., & T. Nasukawa. (2006). *Fully Automatic Lexicon Expanding for Domain-Oriented Sentiment Analysis, EMNLP: Empirical Methods in Natural Language Processing*, http://trl.ibm.com/projects/textmining/takmi/sentiment_analysis_e.htm (accessed February 2014).

Lin, J., & D. Demner-Fushman. (2005). "'Bag of Words' Is Not Enough for Strength of Evidence Classification," *AMIA Annual Symposium Proceedings*, pp. 1031–1032. http://pubmedcen-tral.nih.gov/articlerender.fcgi?artid=1560897 (accessed December 2013).

Mahgoub, H., D. Rösner, N. Ismail, & F. Torkey. (2008). "A Text Mining Technique Using Association Rules Extraction," *International Journal of Computational Intelligence*, 4(1): 21–28.

Manning, C. D., P. Raghavan & H. Schutze. (2008). *Introduction to Information Retrieval*. Cambridge, MA: MIT Press.

Masand, B. M., M. Spiliopoulou, J. Srivastava, & O. R. Zaïane. (2002). "Web Mining for Usage Patterns and Profiles," *SIGKDD Explorations*, 4(2): 125–132.

McKnight, W. (2005, January 1). "Text Data Mining in Business Intelligence," *Information Management Magazine*. http://information-management.com/issues/20050101/1016487-1.html (accessed May 2009).

Mejova, Y. (2009). *Sentiment Analysis: An Overview*, www.cs.uiowa.edu/~ymejova/publications/CompsYelenaMejova.pdf (accessed February 2013).

Miller, T. W. (2005). *Data and Text Mining: A Business Applications Approach*. Upper Saddle River, NJ: Prentice Hall.

Nakov, P., A. Schwartz, B. Wolf, & M. A. Hearst. (2005). "Supporting Annotation Layers for Natural Language Processing." *Proceedings of the ACL*, interactive poster and demonstration sessions. Ann Arbor, MI: Association for Computational Linguistics, pp. 65–68.

Pang, B., & L. Lee. (2008). *Opinion Mining and Sentiment Analysis*, http://dl.acm.org/citation.cfm?id=1596846.

Peterson, E. T. (2008). *The Voice of Customer: Qualitative Data as a Critical Input to Web Site Optimization*, http://foreseeresults.com/Form_Epeterson_WebAnalytics.html (accessed May 2009).

Shatkay, H., et al. (2007). "SherLoc: High-Accuracy Prediction of Protein Subcellular Localization by Integrating Text and Protein Sequence Data," *Bioinformatics*, 23(11): 1410–1417.

StatSoft. (2014). *Statistica Data and Text Miner User Manual*. Tulsa, OK: StatSoft, Inc.

Weng, S. S., & C. K. Liu. (2004) "Using Text Classification and Multiple Concepts to Answer E-mails," *Expert Systems with Applications*, 26(4): 529–543.

Zhou, Y., et al. (2005). "U.S. Domestic Extremist Groups on the Web: Link and Content Analysis," *IEEE Intelligent Systems*, 20(5): 44–51.

7

Big Data Analytics

Using data to understand customers or clients and business operations to sustain (and foster) growth and profitability is an increasingly challenging task for today's enterprises. As more and more data becomes available in various forms and fashion, timely processing of the data with traditional means becomes impractical. This phenomenon is called Big Data, and it is receiving substantial press coverage and drawing increasing interest from both business users and IT professionals. The result is that *Big Data* is becoming an overhyped and overused marketing buzzword.

Big Data means different things to people with different backgrounds and interests. Traditionally, the term Big Data has been used to describe the massive volumes of data analyzed by huge organizations like Google or research science projects at NASA. But for most businesses, it's a relative term: *Big* depends on an organization's size. The point is more about finding new value within and outside conventional data sources. Pushing the boundaries of data analytics uncovers new insights and opportunities, and *big* depends on where you start and how you proceed. Consider this popular description of Big Data: Big Data exceeds the reach of commonly used hardware environments and/or capabilities of software tools to capture, manage, and process it within a tolerable time span for its user population. Big Data has become a popular term to describe the exponential growth, availability, and use of information, both structured and unstructured. Much has been written on the Big Data trend and how it can serve as the basis for innovation, differentiation, and growth.

Where Does Big Data Come From?

A simple answer is "everywhere." The sources of data that were ignored because of the technical limitations are now treated as gold mines. Big Data may come from any number of sources, including blogs, RFID tags, GPS, sensor networks, social networks, Internet-based text documents, Internet search indexes, detailed call records, astronomy, atmospheric science, biology, genomics, nuclear physics, biochemical experiments, medical records, scientific research, military surveillance, photography archives, video archives, and large-scale ecommerce practices.

Figure 7.1 illustrates the sources of Big Data in a three-level diagram. The traditional data sources—mainly business transactions—are illustrated as the first echelon, where the volume, variety, and velocity of the data are moderate to low. The next echelon is the data generated by the Internet and social media. This human-generated data is perhaps the most complicated and potentially most valuable for understanding collective ideas and perceptions of people. The volume, variety, and velocity of data in this echelon are moderate to high. The topmost echelon is machine-generated data. With the automation of data collection systems on many fronts, coupled with the Internet of things (which connects everything to everything else), organizations are now able to collect data at volumes and richness that were unimaginable just a few years ago. All three echelons of data sources create a wealth of information that can significantly improve an organization's capability of solving complex problems and taking advantage of opportunities—if recognized and leveraged properly.

Big Data is not new. What is new is that the definition and the structure of Big Data constantly change. Companies have been storing and analyzing large volumes of data since the advent of the data warehouses in the early 1990s. While terabytes used to be synonymous with Big Data warehouses, now it's petabytes, and the rate of growth in data volumes continues to escalate as organizations seek to store and analyze greater levels of transaction details, as well as Web- and machine-generated data, to gain a better understanding of customer behavior and business drivers. Many—academics and industry analysts and leaders alike—think that "Big Data" is a misnomer. What it says and what it means are not exactly the same. That is, Big Data is not just "big." The sheer volume of the data is only one of many characteristics that are often associated with Big Data, such as variety, velocity, veracity, variability, and value proposition, among others.

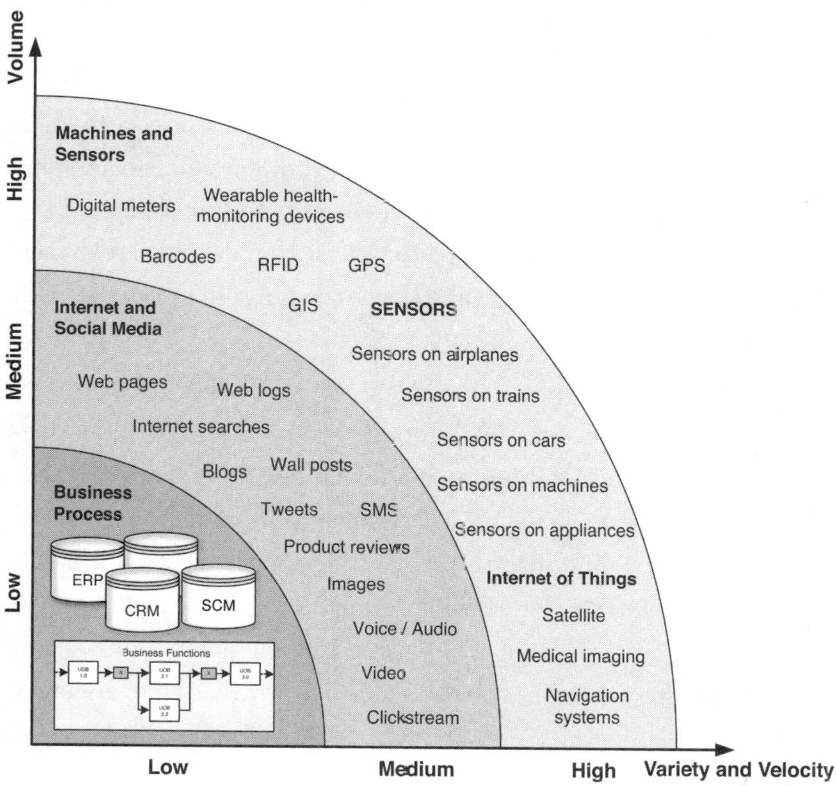

Figure 7.1 The Wide Range of Sources for Big Data

The Vs That Define Big Data

Big Data is typically defined by three *V*s: volume, variety, and velocity. In addition to these three, we see some of the leading Big

Data solution providers adding other Vs, such as veracity (IBM), variability (SAS), and value proposition (literally almost everyone—academics and industry).

Volume

Volume is obviously the most common trait of Big Data. Many factors contributed to the exponential increase in data volume, such as transaction-based data stored through the years, text data constantly streaming in from social media, increasing amounts of sensor data being collected, automatically generated RFID and GPS data, etc. In the past, excessive data volume created storage issues, both technical and financial. But with today's advanced technologies coupled with decreasing storage costs, these issues are no longer significant; instead, other issues emerge, including how to determine relevance amid the large volumes of data and how to create value from data that is deemed to be relevant.

As mentioned before, *big* is a relative term. It changes over time and is perceived differently by different organizations. With the staggering increase in data volume, even the naming of the next Big Data echelon has been challenging. The highest unit of data used to be the petabyte (PB), but now we speak of the zettabyte (ZB), which is a trillion gigabytes (GB) or a billion terabytes (TB). As the volume of data increases, we are having a hard time keeping up with universally accepted naming for the next level. Table 7.1 provides an overview of the size and naming of the modern-day data volumes (Sharda et al. 2014).

Table 7.1 Naming for the Increasing Volumes of Data

Name	Symbol	Value (bytes)
Kilobyte	kB	10^3
Megabyte	MB	10^6
Gigabyte	GB	10^9
Terabyte	TB	10^{12}
Petabyte	PB	10^{15}
Exabyte	EB	10^{18}
Zettabyte	ZB	10^{21}
Yottabyte	YB	10^{24}
Brontobyte*	BB	10^{27}
Gegobyte*	GeB	10^{30}

* Not yet an official SI (International System of Units) name.

Source: en.wikipedia.org/wiki/Petabyte

Consider that an exabyte of data is created on the Internet each day, which equates to 250 million DVDs' worth of information. And the idea of even larger amounts of data—a zettabyte—isn't too far off when it comes to the amount of info traversing the Web in a year. In fact, industry experts are already estimating that we will see 1.3 zettabytes of traffic annually over the Internet by 2016—and soon enough, we might start talking about even bigger volumes. Some of the Big Data scientists often allege that the NSA and FBI have a yottabyte of data on people. To put this measurement in perspective, a yottabyte is the amount of storage on 250 trillion DVDs. A brontobyte, which is not an official SI unit but is apparently recognized by some people in the measurement community, is a 1 followed by 27 zeros. Size of such magnitude can be used to describe the type of sensor data that we will get from the Internet of things in the next decade, if not sooner. A gegobyte is 10^{30}. With respect to where the Big Data comes from, consider the following (Higginbotham, 2012):

- The CERN Large Hadron Collider generates 1 petabyte of data per second.

- Sensors from a Boeing jet engine create 20 terabytes of data every hour.
- Facebook databases take in 500 terabytes of new data per day.
- On YouTube, 72 hours of video are uploaded per minute, translating to 1 terabyte every four minutes.
- The proposed Square Kilometer Array telescope (the proposed biggest telescope in the world) will generate an exabyte of data per day.

From a short historical perspective, in 2009, the world had about 0.8 ZB of data; in 2010, it exceeded the 1 ZB mark; at the end of 2011, the number was 1.8 ZB. IBM has estimated that six or seven years from now, we will have 35 ZB. Though this number is astonishing in size, so are the challenges and opportunities that come with it.

Variety

Data today comes in all types of formats—ranging from traditional databases to hierarchical data stores created by end users and OLAP systems; to text documents, email, and XML; to meter-collected, sensor-captured data; to video, audio, and stock ticker data. By some estimates, 80% to 85% of all organizations' data is in some sort of unstructured or semistructured format (i.e., a format that is not suitable for traditional databases schemas). But there is no denying its value, and hence it must be included in the analyses to support decision making.

Velocity

According to Gartner, a well-known and highly respected technology consultancy company, *velocity* means both how fast data is being produced and how fast the data must be processed (i.e., captured, stored, and analyzed) to meet the need or demand. RFID tags, automated sensors, GPS devices, and smart meters are driving an increasing need to deal with torrents of data in near real time. Velocity is perhaps the most overlooked characteristic of Big Data. Reacting quickly enough to deal with velocity is a challenge to most organizations. For time-sensitive environments, the opportunity cost clock of the data starts ticking the moment the data is created. As time passes, the value proposition of the data degrades and eventually becomes worthless. Whether the subject matter is the health of a patient, the well-being of a traffic system, or the health of an investment portfolio, accessing the data and reacting faster to the circumstances will always create more advantageous outcomes.

In the Big Data storm that we are witnessing today, almost everyone is fixated on at-rest analytics, using optimized software and hardware systems to mine large quantities of variant data sources. Although this is critically important and highly valuable, another class of analytics that is often overlooked is driven by the velocity nature of Big Data; it is called *data stream analytics* or *in-motion analytics*. If done correctly, data stream analytics can be as valuable as—and in some business environments more valuable than—at-rest analytics. Later in this chapter we cover this topic in more detail.

Veracity

IBM is using the term *veracity* as a fourth V to describe Big Data. It refers to conformity to facts—to the accuracy, quality, truthfulness, or trustworthiness of data. Tools and techniques are often used to handle Big Data's veracity by transforming the data into trustworthy insights.

Variability

In addition to the increasing velocities and varieties of data, data flows can be highly inconsistent with periodic peaks. Inconsistency in data flow makes it very difficult to properly and cost effectively develop data infrastructures. If the resources are put in place to handle the peak times, for the rest of the time they will be significantly underutilized. One popular way to handle the problem of variability is to use pooled resources, based on the infrastructure-as-a-service business model. Cloud computing, service oriented architecture, and massively parallel processing make variability a manageable issue for not only large, but also small to medium-sized businesses.

Value Proposition

The excitement around Big Data is about its value proposition. A preconceived notion about Big Data is that it contains (or has a greater potential to contain) more patterns and interesting anomalies than "small" data. Thus, by analyzing large and feature-rich data, organizations can gain greater business value than they could otherwise gain. While users can detect the patterns in small data sets by using simple statistical and machine learning methods or ad hoc query and reporting tools, Big Data means "big" analytics. Big analytics means greater insight and better decision making—something that every organization needs.

Since the exact definition of Big Data is still a matter of ongoing discussion in academic and industrial circles, it is likely that more characteristics (perhaps more Vs) will be added to this list. Regardless of what happens, the importance and value proposition of Big Data are here to stay.

Fundamental Concepts of Big Data

Big Data by itself, regardless of the size, type, or speed of the data, is worthless unless business users do something with it that delivers value to their organizations. That's where "big" analytics comes into the picture. Although organizations have long run reports and dashboards against data warehouses, most have not opened these repositories to in-depth on-demand exploration. This is partly because analysis tools are too complex for the average user and partly because the repositories often do not contain all the data that power users need. But this is about to change (and had already changed for some) in a dramatic fashion, thanks to the new Big Data analytics paradigm.

Along with the value proposition, Big Data has also brought about big challenges for organizations. The traditional means for capturing, storing, and analyzing data are not capable of dealing with Big Data effectively and efficiently. Therefore, new breeds of technologies need to be developed (or purchased, hired, or outsourced) to take on the Big Data challenge. Before making such an investment, organizations should justify the means. Here are some questions that may help shed light on this situation. If any of the following statements are true, then you need to seriously consider embarking on a Big Data journey:

- You can't process the amount of data that you want to process because of the limitations posed by your current platform or environment.
- You want to involve new/contemporary data sources (e.g., social media, RFID, sensory, Web, GPS, textual data) into your analytics platform, but you can't because it does not comply with the data storage schema-defined rows and columns without sacrificing fidelity or the richness of the new data.
- You need to (or want to) integrate data as quickly as possible to be current on your analysis.
- You want to work with a schema-on-demand (as opposed to the predetermined schema used in RDBMSs) data storage paradigm because the nature of the new data may not be known, or there may not be enough time to determine it and develop schema for it.
- The data is arriving so fast at your organization's doorstep that your traditional analytics platform cannot handle it.

As is the case with any other large IT investment, the success in Big Data analytics depends on a number of factors. Figure 7.2 shows a graphical depiction of the most critical success factors (Watson, 2012).

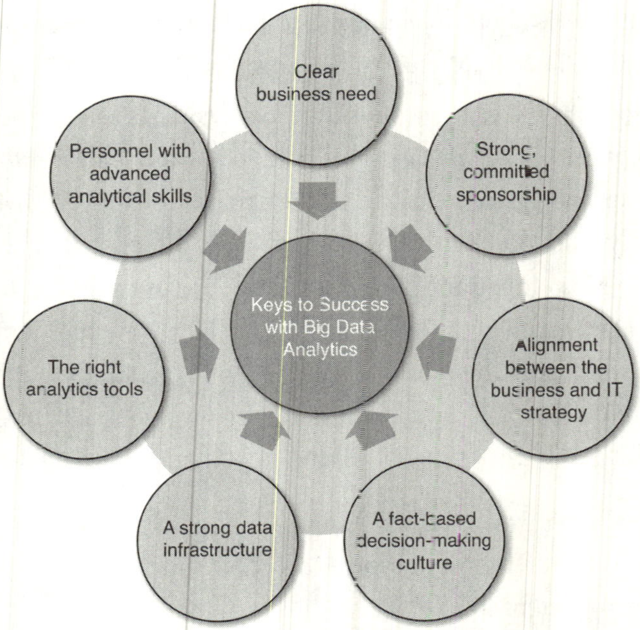

Figure 7.2 Critical Success Factors for Big Data Analytics

Following are among the most critical success factors for Big Data analytics:

- **A clear business need (alignment with the vision and the strategy).** Business investments ought to be made for the good of the business, not for the sake of mere technology advancements. Therefore, the main driver for Big Data analytics should be the needs of the business, at any level: strategic, tactical, or operations.
- **Strong, committed sponsorship (i.e., executive champions).** It is a well-known fact that if you don't have strong, committed executive sponsorship, it is difficult (or even impossible)

to succeed. If the scope is a single or a few analytical applications, the sponsorship can be at the department level. However, if the target is enterprise-wide organizational transformation, which is often the case for Big Data initiatives, sponsorship needs to be at the highest levels, and it needs to be organization-wide.

- **Alignment between the business and IT strategy.** It is essential to make sure that analytics work is always supporting the business strategy—and not the other way around. Analytics should play an enabling role in execution of the business strategy.
- **A fact-based decision-making culture.** In a fact-based decision-making culture, the numbers—rather than intuition, gut feeling, or supposition—drive decision making. There is also a culture of experimentation to see what works and doesn't. To create a fact-based decision-making culture, the senior management needs to
  - Recognize that some people can't or won't adjust
  - Be a vocal supporter
  - Stress that outdated methods must be discontinued
  - Ask to see what analytics went into decisions
  - Link incentives and compensation to desired behaviors
- **A strong data infrastructure.** Data warehouses have provided the data infrastructure for analytics. This infrastructure is changing and being enhanced in the Big Data era with new technologies. Success requires marrying the old with the new for a holistic infrastructure that works synergistically.

As the size and complexity of data increases, the need for more efficient analytical systems is also increasing. In order to keep up with the computational needs of Big Data, a number of new and innovative computational techniques and platforms are being developed. These techniques, collectively called high-performance computing, include the following:

- *In-memory analytics* solves complex problems in near real time with highly accurate insights by allowing analytical computations and Big Data to be processed in-memory and distributed across a dedicated set of nodes.
- *In-database analytics* speeds insights and enables better data governance by performing data integration and analytic functions inside the database so you don't have to move or convert data repeatedly.
- *Grid computing* promotes efficiency, lower cost, and better performance by processing jobs in a shared, centrally managed pool of IT resources.
- *Appliances* bring together hardware and software in a physical unit that is not only fast but also can be scaled on an as-needed basis.

Computational requirement is just a small part of the list of challenges that Big Data imposes on today's enterprises. Following is a list of challenges that business executives have found to have significant impacts on successful implementation of Big Data analytics. When considering Big Data projects and architecture, being mindful of these challenges will make the journey to analytics competency a less stressful one:

- **Data volume.** It's important to be able to capture, store, and process the huge volume of data at an acceptable speed so that the latest information is available to decision makers when they need it.
- **Data integration.** It's important to be able to combine data that is not similar in structure or source and to do so quickly and at reasonable cost.

- **Processing capabilities.** It's important to be able to process the data quickly, as it is captured. Traditional ways of collecting and then processing the data may not work. In many situations, data needs to be analyzed as soon as it is captured to leverage the most value. (This is called stream analytics and is covered later in this chapter.)
- **Data governance.** It's important to be able to keep up with the security, privacy, ownership, and quality issues of Big Data. As the volume, variety (format and source), and velocity of data change, so should the capabilities of governance practices.
- **Skills availability.** Big Data is being harnessed with new tools and is being looked at in different ways. There is a shortage of people (often called data scientists, discussed later in this chapter) with the skills to do the job.
- **Solution cost.** Since Big Data has opened up a world of possible business improvements, there is a great deal of experimentation and discovery taking place to determine the patterns that matter and the insights that turn to value. To ensure a positive ROI on a Big Data project, therefore, it is crucial to reduce the cost of the solutions used to find that value.

The challenges are real, but so is the value proposition of Big Data analytics. Anything that business analytics leaders can do to help prove the value of new data sources to the business will move the organization beyond experimenting and exploring Big Data to adapting and embracing it as a differentiator. There is nothing wrong with exploration, but ultimately the value of Big Data comes from putting insights into action.

The Business Problems That Big Data Analytics Addresses

The top business problems addressed by Big Data overall are process efficiency and cost reduction, as well as enhanced customer experience, but different priorities emerge when it is looked at by industry. Process efficiency and cost reduction are common business problems that can be addressed by analyzing Big Data, and these are perhaps among the top-ranked problems that can be addressed with Big Data analytics for the manufacturing, government, energy and utilities, communications and media, transport, and healthcare sectors. Enhanced customer experience may be at the top of the list of problems addressed by insurance companies and retailers. Risk management usually is at the top of the list for companies in banking and education. Here is a list of problems that can be addressed using Big Data analytics:

- Process efficiency and cost reduction
- Brand management
- Revenue maximization, cross-selling, and up-selling
- Enhanced customer experience
- Churn identification and customer recruiting
- Improved customer service
- Identification of new products and market opportunities
- Risk management
- Regulatory compliance
- Enhanced security capabilities

Big Data Technologies

There are a number of technologies to processing and analyzing Big Data, but most have some common characteristics (Kelly, 2012).

Namely, they take advantage of commodity hardware to enable scale-out and parallel-processing techniques; they employ nonrelational data storage capabilities in order to process unstructured and semi-structured data; and they apply advanced analytics and data visualization technology to Big Data to convey insights to end users. There are three Big Data technologies that stand out as likely to transform the business analytics and data management markets: MapReduce, Hadoop, and NoSQL. The following sections examine these three technologies.

MapReduce

MapReduce is a technique popularized by Google that distributes the processing of very large multistructured data files across a large cluster of machines. High performance is achieved by breaking the processing into small units of work that can be run in parallel across the hundreds—potentially thousands—of nodes in the cluster. In their seminal paper on MapReduce, Dean and Ghemawat (2004) said:

> MapReduce is a programming model and an associated implementation for processing and generating large data sets. Programs written in this functional style are automatically parallelized and executed on a large cluster of commodity machines. This allows programmers without any experience with parallel and distributed systems to easily utilize the resources of a large distributed system.

The key point to note from this quote is that MapReduce is a programming model, not a programming language. That is, it is designed to be used by programmers rather than by business users.

To see how MapReduce works, let's look at an example. In the colored square counter in Figure 7.3, the input to the MapReduce process is a set of colored squares. The objective is to count the number of squares of each color. The programmer in this example is responsible for coding the map and reduce programs; the remainder of the processing is handled by the software system implementing the MapReduce programming model.

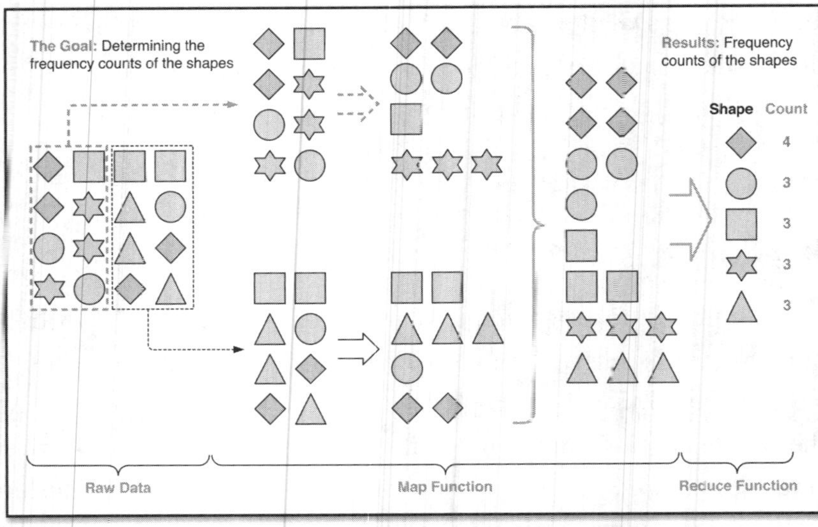

Figure 7.3 A Graphical Depiction of the MapReduce Process

The MapReduce system first reads the input file and splits it into multiple pieces. In this example, there are two splits, but in a real-life scenario, the number of splits would typically be much higher. These splits are then processed by multiple map programs running in parallel on the nodes of the cluster. The role of each map program in this case is to group the data in a split by color. The MapReduce system then takes the output from each map program and merges (i.e., shuffles and sorts) the results for input to the reduce program, which calculates the sum of the number of squares of each color. In this example, only one copy of the reduce program is used, but there may be more in practice. To optimize performance, programmers can provide their own shuffle and sort program, and they can also deploy a combiner that combines local map output files to reduce the number of output files that have to be remotely accessed across the cluster by the shuffling and sorting step.

Why Use MapReduce?

MapReduce aids organizations in processing and analyzing large volumes of multistructured data. Application examples include indexing and search, graph analysis, text analysis, machine learning, and data transformation. These types of applications are often difficult to implement using the standard SQL employed by relational DBMSs.

The procedural nature of MapReduce makes it easy for skilled programmers to understand. It also has the advantage of not requiring developers to be concerned with implementing parallel computing; the system transparently handles that.

Although MapReduce is designed for programmers, nonprogrammers can exploit the value of prebuilt MapReduce applications and function libraries. Both commercial and open source MapReduce libraries are available that provide a wide range of analytic capabilities. Apache Mahout, for example, is an open source machine learning library of algorithms for clustering, classification, and batch-based collaborative filtering that are implemented using MapReduce.

Hadoop

Hadoop is an open source framework for processing, storing, and analyzing massive amounts of distributed, unstructured data. Originally created by Doug Cutting at Yahoo!, Hadoop was inspired by

MapReduce, a user-defined function developed by Google in the early 2000s for indexing the Web. Hadoop was designed to handle petabytes and exabytes of data, distributed over multiple nodes in parallel.

Hadoop clusters run on inexpensive commodity hardware, so projects can scale out without breaking the bank. Hadoop is now a project of the Apache Software Foundation, where hundreds of contributors continuously improve the core technology.

How Hadoop Works

Fundamentally, rather than bang away at one huge block of data with a single machine, Hadoop breaks up Big Data into multiple parts so all the parts can be processed and analyzed at the same time. A client accesses unstructured and semistructured data from sources including log files, social media feeds, and internal data stores. It breaks the data up into parts, which are then loaded into a file system made up of multiple nodes running on commodity hardware. The default file store in Hadoop is the Hadoop Distributed File System (HDFS). File systems such as HDFS are adept at storing large volumes of unstructured and semistructured data because they do not require data to be organized into relational rows and columns. Each part is replicated multiple times and loaded into the file system so that if a node fails, another node has a copy of the data contained on the failed node. A name node acts as facilitator, communicating back to the client information such as which nodes are available, where in the cluster certain data resides, and which nodes have failed.

Once the data is loaded into the cluster, it is ready to be analyzed via the MapReduce framework. The client submits a map job—usually a query written in Java—to one of the nodes in the cluster known as the job tracker. The job tracker refers to the name node to determine which data it needs to access to complete the job and where in the cluster that data is located. Once this is determined, the job tracker submits the query to the relevant nodes. Rather than bring all the data back into a central location for processing, processing occurs at each node simultaneously, or in parallel. This is an essential characteristic of Hadoop.

When each node has finished processing its given job, it stores the results. The client initiates a reduce job through the job tracker, in which results of the map phase stored locally on individual nodes are aggregated to determine the "answer" to the original query and are then loaded onto another node in the cluster. The client accesses these results, which can then be loaded into one of a number of analytic environments for analysis. The MapReduce phase has now been completed.

After the MapReduce phase, the processed data is ready for further analysis by data scientists and others with advanced data analytics skills. Data scientists can manipulate and analyze the data, using a number of tools for a number of uses, including to search for hidden insights and patterns or to use as the foundation to build user-facing analytic applications. The data can also be modeled and transferred from Hadoop clusters into existing relational databases, data warehouses, and other traditional IT systems for further analysis and/or to support transactional processing.

Hadoop Technical Components

A Hadoop "stack" is made up of a number of components, including the following:

- **Hadoop Distributed File System (HDFS).** This is the default storage layer in any given Hadoop cluster.
- **Name node.** This is the node in a Hadoop cluster that provides the client information on where in the cluster particular data is stored and whether any nodes fail.

- **Secondary node.** A backup to the name node, the secondary node periodically replicates and stores data from the name node in case it fails.
- **Job tracker.** This is the node in a Hadoop cluster that initiates and coordinates MapReduce jobs or the processing of the data.
- **Slave nodes.** The grunts of any Hadoop cluster, slave nodes store data and take direction to process it from the job tracker.

In addition, the Hadoop ecosystem is made up of a number of complimentary subprojects. NoSQL data stores like Cassandra and HBase are used to store the results of MapReduce jobs in Hadoop. In addition to Java, some MapReduce jobs and other Hadoop functions are written in Pig, an open source language designed specifically for Hadoop. Hive is an open source data warehouse originally developed by Facebook that allows for analytic modeling within Hadoop. A rich set of Hadoop-related projects and supporting tools/platforms can be found in Sharda et al. (2014).

Hadoop Pros and Cons

The main benefit of Hadoop is that it allows enterprises to process and analyze large volumes of unstructured and semistructured data, heretofore inaccessible to them, in a cost- and time-effective manner. Because Hadoop clusters can scale to petabytes and even exabytes of data, enterprises no longer need to rely on sample data sets but can process and analyze *all* relevant data. Data scientists can apply an iterative approach to analysis, continually refining and testing queries to uncover previously unknown insights. Also, getting started with Hadoop doesn't cost much. Developers can download the Apache Hadoop distribution for free and begin experimenting with Hadoop in less than a day.

The downside to Hadoop and its myriad components is that they are immature and still developing. As with any other young, raw technology, implementing and managing Hadoop clusters and performing advanced analytics on large volumes of unstructured data requires significant expertise, skill, and training. Unfortunately, there is currently a dearth of Hadoop developers and data scientists available, which means it's impractical for many enterprises to maintain and take advantage of complex Hadoop clusters. Further, as Hadoop's myriad components are improved by the community and as new components are created, there is (as there is with any immature open source technology/approach) a risk of forking. Finally, Hadoop is a batch-oriented framework, which means it does not support real-time data processing and analysis.

The good news is that some of the brightest minds in IT are contributing to the Apache Hadoop project, and a new generation of Hadoop developers and data scientists is coming of age. As a result, the technology is advancing rapidly, becoming both more powerful and easier to implement and manage. An ecosystems of vendors, both Hadoop-focused startups like Cloudera and Hortonworks and well-worn IT stalwarts like IBM and Microsoft, are working to offer commercial, enterprise-ready Hadoop distributions, tools, and services to make deploying and managing the technology a practical reality for traditional enterprises. Other bleeding-edge startups are working to perfect NoSQL (which stands for Not Just SQL) data stores capable of delivering near-real-time insights in conjunction with Hadoop.

A Few Demystifying Facts About Hadoop

Although Hadoop and related technologies, such as MapReduce and Hive, have been around for over five years now, many people still have several misconceptions about them. The following list of 10 facts intends to clarify what Hadoop is and does relative to BI, as well as in which business and technology situations Hadoop-based BI, data warehousing, and analytics can be useful (Russom, 2013):

- **Fact 1: Hadoop consists of multiple products.** We talk about Hadoop as if it's one monolithic thing, but it's actually a family of open source products and technologies *overseen* by the Apache Software Foundation (ASF). (Some Hadoop products are also available via vendor distributions; more on that in the next point.)

The Apache Hadoop library includes (in BI priority order) the Hadoop Distributed File System (HDFS), MapReduce, Hive, HBase, Pig, Zookeeper, Flume, Sqoop, Oozie, Hue, and so on. You can combine these in various ways, but HDFS and MapReduce (perhaps with HBase and Hive) constitute a useful technology stack for applications in BI, DWs, and analytics.

- **Fact 2: Hadoop is open source but available from vendors, too.** Apache Hadoop's open source software library is available from ASF at www.apache.org. For users desiring a more enterprise-ready package, a few vendors now offer Hadoop distributions that include additional administrative tools and technical support.
- **Fact 3: Hadoop is an ecosystem, not a single product.** In addition to products from Apache, the extended Hadoop ecosystem includes a growing list of vendor products that integrate with or expand Hadoop technologies. One minute on your favorite search engine will reveal these.
- **Fact 4: HDFS is a file system, not a DBMS.** Hadoop is primarily a distributed file system and lacks capabilities associated with a DBMS, such as indexing, random access to data, and support for SQL. That's okay because HDFS does things DBMSs cannot do.
- **Fact 5: Hive resembles SQL but is not standard SQL.** Many of us are handcuffed to SQL because we know it well, and our tools demand it. People who know SQL can quickly learn to hand-code Hive, but that doesn't solve compatibility issues with SQL-based tools. The Data Warehouse Institute (TDWI) feels that over time, Hadoop products will support standard SQL, so this issue will soon be moot.

- **Fact 6: Hadoop and MapReduce are related but don't require each other.** Developers at Google developed MapReduce before HDFS existed, and some variations of MapReduce work with a variety of storage technologies, including HDFS, other file systems, and some DBMSs.
- **Fact 7: MapReduce provides control for analytics, not analytics per se.** MapReduce is a general-purpose execution engine that handles the complexities of network communication, parallel programming, and fault tolerance for any kind of application that you can hand-code—not just analytics.
- **Fact 8: Hadoop is about data diversity, not just data volume.** Theoretically, HDFS can manage the storage and access of any data type, as long as you can put the data in a file and copy that file into HDFS. As outrageously simplistic as that sounds, it's largely true, and it's exactly what brings many users to Apache HDFS.
- **Fact 9: Hadoop complements a DW; it's rarely a replacement.** Most organizations have designed their DW for structured, relational data, which makes it difficult to wring BI value from unstructured and semistructured data. Hadoop promises to complement DWs by handling the multistructured data types that most DWs can't handle.
- **Fact 10: Hadoop enables many types of analytics, not just Web analytics.** Hadoop gets a lot of press about how Internet companies use it for analyzing Web logs and other Web data. But other use cases exist. For example, consider the Big Data coming from sensory devices, such as robotics in manufacturing, RFID in retail, or grid monitoring in utilities. Older analytic applications that need large data samples—such as customer-base segmentation, fraud detection, and risk analysis—can benefit from the additional Big Data managed by Hadoop. Likewise, Hadoop's additional data can expand 360-degree views to create a more complete and granular view.

NoSQL

A related new style of database called NoSQL (which stands for Not Only SQL) has emerged to, like Hadoop, process large volumes of multistructured data. However, whereas Hadoop is adept at supporting large-scale, batch-style historical analysis, NoSQL databases are aimed, for the most part (though there are some important exceptions), at serving up discrete data stored among large volumes of multistructured data to end-user and automated Big Data applications. This capability is sorely lacking from relational database technology, which simply can't maintain needed application performance levels at a Big Data scale.

In some cases, NoSQL and Hadoop work in conjunction. HBase, for example, is a popular NoSQL database modeled after Google BigTable that is often deployed on top of HDFS to provide low-latency, quick lookups in Hadoop. The downside of most NoSQL databases today is that they have traded ACID (atomicity, consistency, isolation, durability) compliance for performance and scalability. Many also lack mature management and monitoring tools. Both these shortcomings are in the process of being overcome by both the open source NoSQL communities and a handful of vendors that are attempting to commercialize the various NoSQL databases. NoSQL databases currently available include HBase, Cassandra, MongoDB, Accumulo, Riak, CouchDB, and DynamoDB, among others.

Data Scientists

Data scientist is a role or a job frequently associated with Big Data or data science. In a very short time, it has become one of the most sought-out roles in the marketplace. In an article published in the October 2012 issue of *Harvard Business Review*, authors Thomas H. Davenport and D. J. Patil called data scientist "the sexiest job of the 21st century." In their article, they specified data scientists' most basic, universal skill as the ability to write code (in the latest Big Data languages and platforms). Although this may be less true in the near future, when many more people will have the title "data scientist" on their business cards, at this time it seems to be the most fundamental skill required of data scientists. A more enduring skill will be the need for data scientists to communicate in language that all their stakeholders understand—and to demonstrate the special skills involved in storytelling with data, whether verbally, visually, or—ideally—both (Davenport & Patil, 2012).

Data scientists use a combination of business and technical skills to *investigate* Big Data, looking for ways to improve current business analytics practices (from descriptive to predictive and prescriptive) and hence to improve decisions for new business opportunities. One of the biggest differences between a data scientist and a business intelligence user—such as a business analyst—is that a data scientist investigates and looks for new possibilities, while a BI user analyzes existing business situations and operations.

One of the dominant traits expected of data scientists is an intense curiosity—a desire to go beneath the surface of a problem, find the questions at its heart, and distill them into a very clear set of hypotheses that can be tested. This often entails the associative thinking that characterizes the most creative scientists in any field. For example, one data scientist studying a fraud problem realized that it was analogous to a type of DNA sequencing problem (Davenport & Patil, 2012). By bringing together those disparate worlds, he and his team were able to craft a solution that dramatically reduced fraud losses.

Where Do Data Scientists Come From?

Although there still is disagreement about the use of *science* in the name, data science is becoming less of a controversial issue. Real scientists use tools made by other scientists, or they make them if they don't exist, as a means to expand knowledge. That is exactly what data scientists do. Experimental physicists, for example, have to design equipment, gather data, conduct multiple experiments to discover knowledge, and communicate their results. Even though they may not be wearing white coats and may not be living in a sterile lab environment, data scientists have a similar role to experimental physicists in that they use creative tools and techniques to turn data into actionable information for others to use for better decision making.

There is no consensus on what educational background a data scientist has to have. The usual suspects, like MS or PhD in computer science, MIS, industrial engineering, or, recently, analytics, may be necessary but not sufficient to call someone a data scientist. One of the most sought-out characteristics of a data scientist is expertise in both technical and business application domains. In that sense, it somewhat resembles the professional engineer (PE) or project management professional (PMP) roles, where experience is valued as much as (if not more so than) technical skills and educational background. It would not be a huge surprise to see in the next few years a certification specifically designed for data scientists.

Because it is a profession for a field that is still being defined and many of whose practices are still experimental and far from being standardized, companies are overly sensitive about the experience dimension of data scientists. As the profession matures and practices are standardized, experience will be less of an issue in the definition of a data scientist. Today, companies are looking for people who have extensive experience working with complex data and have had good luck recruiting among those with educational and work backgrounds in the physical or social sciences. Some of the best and brightest data scientists have been PhDs in esoteric fields like ecology and systems biology (Davenport & Patil, 2012). Even though there is no consensus on where data scientists come from, there is a common understanding of what skills and qualities they are expected to possess. Figure 7.4 shows a high-level graphical illustration of skills that a data scientist needs.

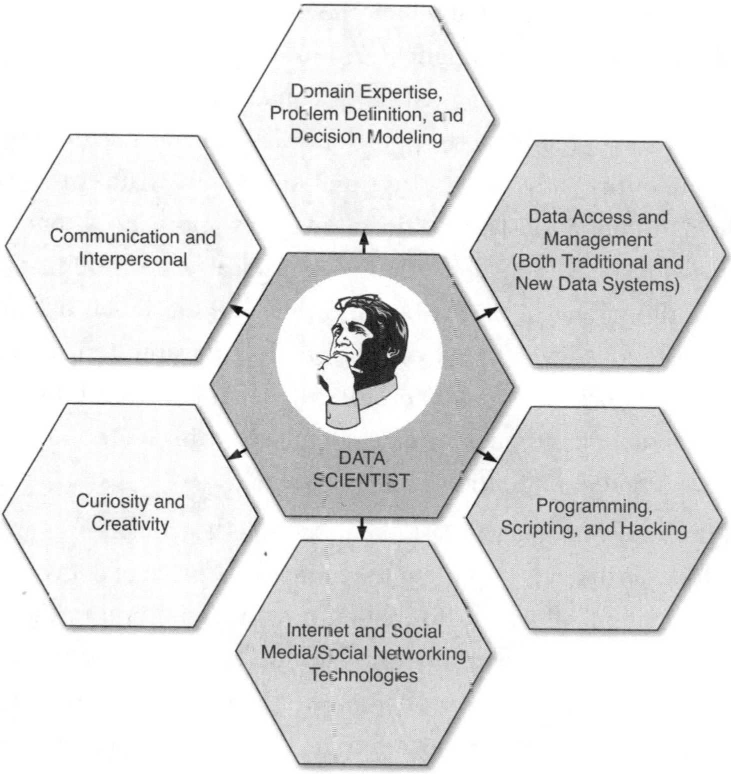

Figure 7.4 Skills That a Data Scientist Needs

Data scientists need to have soft skills such as creativity, curiosity, communication/interpersonal, domain expertise, problem definition, and managerial skills (shown on the left side of Figure 7.4). They also need sound technical skills such as data manipulation, programming/hacking/scripting, and Internet and social media/networking technologies (shown on the right side of the figure).

Big Data and Stream Analytics

Along with volume and variety, as we have seen earlier in this chapter, one of the key characteristics defining Big Data is velocity, which refers to the speed at which the data is created and streamed into the analytics environment. Organizations are looking for new means to process streaming data as it comes in to react quickly and accurately to problems and opportunities to please customers and gain competitive advantage. In situations where data streams in rapidly and continuously, traditional analytics approaches that work with previously accumulated data (i.e., data at arrest) often either arrive at the wrong decisions because they use too much of out-of-context data, or they arrive at the correct decisions but too late to be of any use to organization. Therefore, it is critical for a number of business situations to analyze data soon after it is created and/or as soon as it streamed into the analytics system.

A presumption that the vast majority of modern businesses are living by today is that it is important to record every piece of data because it might contain valuable information now or sometime in the near future. However, as the number of data sources increases, the store-everything approach becomes harder and harder to do and, in some cases, even infeasible. In fact, despite the technological advances, current total storage capacity lags far behind the digital information being generated in the world. Moreover, in the constantly changing business environment, real-time detection of meaningful changes in data as well as of complex pattern variations within a given short time window are essential in order to come up with actions that better fit with the new environment. These facts become the main triggers for a paradigm called *stream analytics*. The stream analytics paradigm was born as an answer to challenges such as the unbounded flows of data that cannot be permanently stored in order to be subsequently analyzed, in a timely and efficient manner, and complex pattern variations that need to be detected and acted upon as soon as they happen.

Stream analytics (also called *data-in-motion analytics* and *real-time data analytics*) is a term commonly used for the analytic process of extracting actionable information from continuously streaming data. A *stream* can be defined as a continuous sequence of data elements (Zikopoulos et al., 2013). The data elements in a stream are often called *tuples*. In a relational database sense, a tuple is similar to a row of data (i.e., a record, an object, or an instance). However in the context of semistructured or unstructured data, a tuple is an abstraction that represents a package of data, which can be characterized as a set of attributes for a given object. If a tuple by itself is not sufficiently informative for analysis, a correlation or another collective relationship among tuples is needed, and then a window of data that includes a set of tuples is used. A window of data is a finite number/sequence of tuples, and the window is continuously updated as new data becomes available. The size of a window is determined based on the system being analyzed. Stream analytics is becoming increasingly more popular because of two things: First, time-to-action has become a decreasing value, and second, we have the technological means to capture and process the data while it is being created.

Some of the most impactful applications of stream analytics have been developed in the energy industry, specifically for smart grid (electric power supply chain) systems. The new smart grids are capable of not only real-time creation and processing of multiple streams of data in order to determine optimal power distribution to fulfill real customer needs but also generate accurate short-term predictions aiming at covering unexpected demand and renewable energy generation peaks.

Figure 7.5 shows a depiction of a generic use case for streaming analytics in the energy industry (a typical smart grid application). The goal is to accurately predict electricity demand and production in real time by using streaming data that is coming from smart meters, production system sensors, and meteorological models. The ability to predict near-future consumption and production trends and detect anomalies in real time can be used to optimize supply decisions (e.g., how much to produce, what sources of production to use, how to optimally adjust production capacities) as well as to adjust smart meters to regulate consumption and favorable energy pricing.

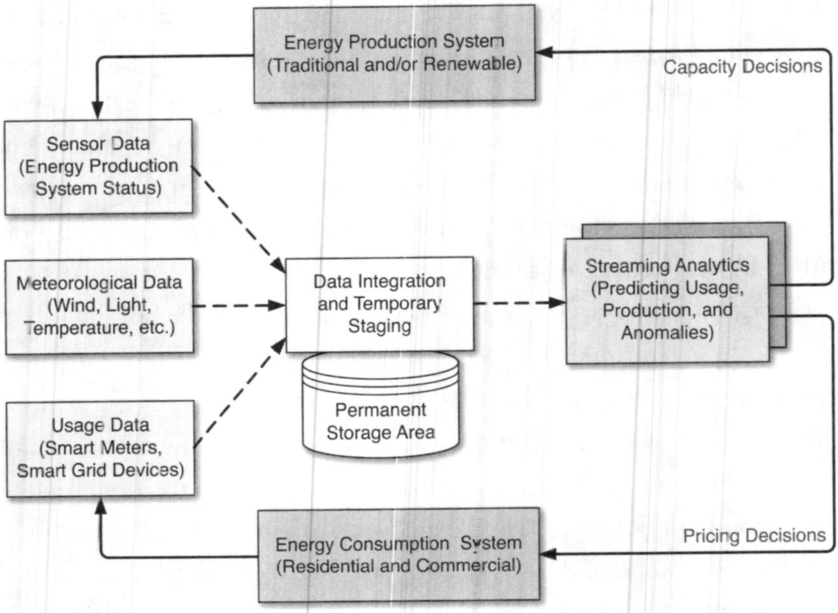

Figure 7.5 A Use Case of Streaming Analytics in the Energy Industry

Data Stream Mining

Data stream mining, as an enabling technology for stream analytics, is the process of extracting novel patterns and knowledge structures from continuous, rapid data records. As we have seen in this book, traditional data mining methods require data to be collected and organized in a proper file format and then processed in a recursive manner to learn the underlying patterns. In contrast, a data stream is a continuous flow of an ordered sequence of instances that in many applications of data stream mining can be read/processed only once or a small number of times, using limited computing and storage capabilities. Examples of data streams include sensor data, computer network traffic, phone conversations, ATM transactions, Web searches, and financial data. Data stream mining can be considered a subfield of data mining, machine learning, and knowledge discovery.

In many data stream mining applications, the goal is to predict the class or value of new instances in the data stream, given some knowledge about the class membership or values of previous instances in the data stream. Specialized machine learning techniques (mostly derivative of traditional machine learning techniques) can be used to learn this prediction task from labeled examples in an automated fashion. An example of such a prediction method has been developed by Delen et al. (2005), who gradually built and refined a decision tree model by using a subset of the data at a time.

Application Example: Big Data for Political Campaigns

One of the application areas where Big Data and analytics promise to make a big difference is the field of politics. Experiences from the recent presidential elections have illustrated the power of Big Data and analytics to acquire and energize millions of volunteers (in the form of a modern-era grassroots movement) to not only raise hundreds of millions of dollars for the election campaign but to optimally organize and mobilize potential voters to get out and vote in large numbers. Clearly, the 2008 and 2012 U.S. presidential elections made a mark on the political arena with the creative use of Big Data and analytics to improve the chances of winning. Figure 7.6 illustrates a graphical depiction of the analytical process of converting a wide variety of data to the ingredients for winning an election.

As Figure 7.6 illustrates, data is the source of information; the richer and deeper it is, the better and more relevant the insights. The main characteristics of Big Data—volume, variety, and velocity (the three *V*s)—readily apply to the kind of data that is used for political campaigns. In addition to the structured data (e.g., detailed records of previous campaigns, census data, market research and poll data), vast volumes and verity of social media (e.g., tweets at Twitter, Facebook wall posts, blog posts), and Web data (Web pages, news articles, newsgroups) are used to learn more about voters and obtain deeper insights to enforce or change them. Often, the search and browsing histories of individuals are captured and made available to customers (political analysts), who can use such data for better insight and behavioral targeting. If done correctly, Big Data and analytics can provide invaluable information to manage political campaigns better than ever before.

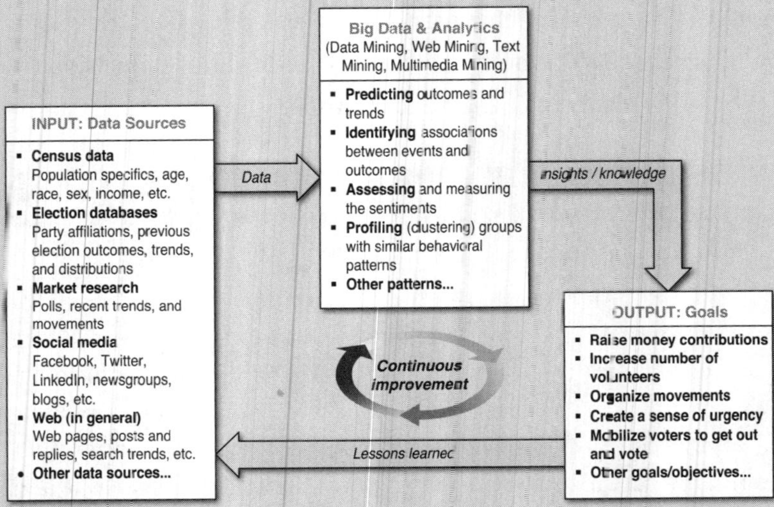

Figure 7.6 Leveraging Big Data and Analytics for Political Campaigns

From predicting the election outcomes to targeting potential voters and donors, Big Data and analytics has a lot to offer to modern election campaigns. In fact, it has changed the way presidential election campaigns are run. In the 2008 and 2012 presidential elections, the major political parties (Republican and Democrat) employed social media and data-driven analytics for more effective and efficient campaigns, but as many agree, the Democrats clearly had the competitive advantage (Issenberg, 2012). Obama's 2012 data- and analytics-driven operation was far more sophisticated and more efficient than its much-heralded 2008 process, which was primarily social media driven. In the 2012 campaign, hundreds of analysts applied advanced analytics on very large and diverse data sources to pinpoint exactly who to target, for what reason, and with what message—on a continuous basis. Compared to 2008, they had more expertise, hardware, software, data (e.g., Facebook and Twitter were orders of magnitude bigger in 2012 than they were in 2008), and computational resources to go over and beyond what they had accomplished previously (Shen, 2013). Before the 2012 election, a *Politico* reporter claimed that Obama had a data advantage and went on to say that the depth and breadth of the campaign's digital operation, from political and demographic data mining to voter sentiment and behavioral analysis, reached beyond anything politics had ever seen (Romano, 2012).

According to Shen (2013), the real winner of the 2012 election was analytics. While most people, including so-called political experts (who often rely on gut feelings and experiences), thought the 2012 presidential election would be very close, a number of analysts, based on their data-driven analytical models, predicted that Obama would win easily, with close to 99% certainty. For example, Nate Silver at FiveThirtyEight, a popular political blog published by *The New York Times*, predicted not only that Obama would win but by exactly how much. Simon Jackman, professor of political science at Stanford University, accurately predicted that Obama would win 332 electoral votes and that North Carolina and Indiana would be the only two states that Obama won in 2008 that would fall to Romney.

In short, Big Data and analytics have become a critical part of political campaigns. The use and expertise gap between the party lines may disappear, but the importance of analytical capabilities will continue to evolve in the foreseeable future.

Sources: Compiled from G. Shen, "Big Data, Analytics and Elections," *INFORMS Analytics Magazine*, January–February 2013; L. Romano, "Obama's Data Advantage," *Politico*, June 9, 2012; M. Scherer, "Inside the Secret World of the Data Crunchers Who Helped Obama Win," *Time*, November 7, 2012; S. Issenberg, "Obama Does It Better" (from "Victory Lab: The New Science of Winning Campaigns"), *Slate*, October 29, 2012; and D. A. Samuelson, "Analytics: Key to Obama's Victory," *ORMS Today*, February 2013, pp. 20–24.

References

Davenport, T. H., & D. J. Patil. (2012, October). "Data Scientist." *Harvard Business Review*, pp. 70–76.

Dean, J., & S. Ghemawat. (2004). *MapReduce: Simplified Data Processing on Large Clusters*. http://research.google.com/archive/mapreduce.html (accessed May 2014).

Delen, D., M. Kletke, & J. Kim. (2005). "A Scalable Classification Algorithm for Very Large Datasets," *Journal of Information and Knowledge Management*, 4(2): 83–94.

Higginbotham, S. (2012). *As Data Gets Bigger, What Comes After a Yottabyte?* http://gigaom.com/2012/10/30/as-data-gets-bigger-what-comes-after-a-yottabyte (accessed June 2014).

Issenberg, S. (2012, October 29). "Obama Does It Better." *Slate*.

Kelly, L. (2012). *Big Data: Hadoop, Business Analytics and Beyond*. wikibon.org/wiki/v/Big_Data:_Hadoop,_Business_Analytics_and_Beyond (accessed June 2014).

Romano, L. (2012, June 9). "Obama's Data Advantage," *Politico*.

Russom, P. (2013). "Busting 10 Myths About Hadoop," *Best of Business Intelligence*, 10: 45–46.

Samuelson, D. A. (2013, February). "Analytics: Key to Obama's Victory," *ORMS Today*, pp. 20–24.

Sharda, R., D. Delen, & E. Turban. (2014). *Business Intelligence and Analytics: Systems for Decision Support*. New York: Prentice Hall.

Shen, G. (2013, January–February). "Big Data, Analytics and Elections," *INFORMS Analytics Magazine*.

Watson, H. (2012). "The Requirements for Being an Analytics-Based Organization," *Business Intelligence Journal*, 17(2): 42–44.

Watson, H., R. Sharda, & D. Schrader. (2012). "Big Data and How to Teach It," Workshop at AMCIS 2012, Seattle.

Zikopoulos, P., et al. (2013). *Harness the Power of Big Data*. New York: McGraw-Hill.

Index

Numerics

A

D

E

Q-R

S